WHAT WE TALK ABOUT WHEN WE TALK ABOUT
INNOVATION

The advent of smart technology explained to people in a hurry

ÁLVARO PÉREZ

WHAT WE TALK ABOUT WHEN WE TALK ABOUT INNOVATION

ÁLVARO PÉREZ

This work is subject to Creative Commons Attribution-NonCommercial 4.0 International license.

To view a copy of this license, visit http://creativecommons.org/licenses/by-nc/4.0/

ISBN of the print edition: 979-8-65573-431-9.

This work has been registered under a CC license at safecreative.org with identifier 2006214486720.

Original title: *De qué hablamos cuando hablamos de innovar-*
First edition for the English version: June 2020.
Edition and proofreading for the English version: Sarah Pila and Emma Moylan.
Cover image credit: Freepik.com
Fonts used: Cochin, Gagalin (by Iordanis Passas) and Bristol (by Jovanny Lemonad).

To all who believe in a better world.

INDEX

- Foreword 9
- Theory 11
  - The birth of tragedy 13
  - We are not alone 41
  - A robot took my job away 97
  - Digital metamorphosis 113
- Practice 135
  - How to innovate 137
  - Start an innovation factory 169
  - Common walls 193
  - Agile 223
- Epilogue 241
  - Ethics and politics 243

FOREWORD

This book summarizes what innovation is, its history, and the most important technologies that will affect us in the coming years. It focuses on intangible technologies more than manufacturing, but the cast is broad and varied: software tools, agile methodologies, business models, artificial intelligence, patents, open innovation, public subsidies, and more. Space is limited, and I have made sure of setting myself a discreet goal. The book covers 270 pages. It can be read in under five hours. Consequently, its depth is limited. Because its scope is wide and it tries to adapt to all audiences, the level of technical detail is insufficient for its direct implementation. But it should generate curiosity to continue investigating each of the topics.

The book is divided into two parts.

The first part explores technological evolution to this day and introduces new forms of business that digitization has enabled in this century. It tackles basic concepts, such as the switch from analog to digital media or the decrease in capital needs to create a company. In recent years, the world has witnessed some revolutionary phenomena: the connected crowd, platform models, the democratization of technologies like big data or the Cloud. How does this affect the way companies innovate and people reinvent themselves? How can the new generations take advantage of it? Everything ancient faces the same challenge, how to transform?

The second part is pragmatic and focused on the business world. It is designed for those who work in a corporation and can put the key findings of this book into practice, but still interesting for anyone who wants to understand the path of organizations towards the digital world.

Innovation is key to the good performance and sustainability of organizations. As machines continue to increase their presence at work, fewer employees will be required. It is essential to put together the right mix of strategy, structure, culture, and processes, under the understanding of new tools, methodologies, and technological trends.

I hope you enjoy it.

THEORY

THE BIRTH OF TRAGEDY

1

"If I have seen further, it is by standing on the shoulders of Giants."
—Isaac Newton, 1643–1727

Then we came.

According to Greek tradition, Prometheus created man from clay and gave him fire, a symbol of skill and self-sufficiency, for which he suffered the wrath and punishment of Zeus.

The myth of human creation from a certain material, in particular clay, is repeated in countless other creeds. In Egypt, it is Jnum, in the Mesopotamian Gilgamesh, Enkidu. The same occurs in the American religions prior to the conquest, both the northern natives and the Incas with Huiracocha. The Christian god made the woman from the rib of the man, who comes from the dust, according to Genesis 2:7, a book taken from the Jews. In Quran 15:26, Allah makes Adam from clay. We also find the theft of fire throughout various mythological traditions. This gives us an incipient and paradigmatic clue of what it means to innovate: to build on the shoulders of the former.

But something interesting appears in Plato's version:

> Then Prometheus, in his perplexity as to what preservation he could devise for man, stole from Hephaestus and Athena wisdom in the arts together with fire—since by no means without fire could it be acquired or helpfully used by any—and he handed it there and then as a gift to man
> —Plato, Protagoras, 320d–322a.

Prometheus takes more than fire: He steals the technical skill, the ability to do things for oneself, for which we need fire as a tool. The value is not in the technology, but in the talent to create. In Hesiod's version, older than Plato's, Zeus feels enraged by the Promethean gift and sends Pandora—forged by Hephaestus from clay—to Earth, the first woman among men. A carrier, like Eve, of all our misfortunes. In all versions, we think of the plunder of the Titan Prometheus as a challenge to the power embodied in Zeus. Have human beings developed the arts and techniques from a theft that the gods wanted to prevent? They have innovated and devised tools that have made their existence easier and, sometimes, worse. What if Zeus was right? What if that act was as cursed for our species as the bite that expelled us from Eden? Perhaps this wild hunger for innovation will lead us to extinction. Perhaps human greed will make our planet vanish and, as Olympus suspected, it would have been better to stay away from fire.

Truth is, the history of man reflects a complex aggregation system of practices and methods, from the Stone and Bronze Ages to the present day. Its existence is closely linked to the quest for fresh challenges. If there is a meaning to life, man has not found it. To fight against that uncertainty, he scans the future, explores unprecedented lands on and off the planet, comes up with original methods of achieving the same old things.

Archeological remains show that *Homo erectus* knew about fire a million and a half years ago. They understood how to handle it, but not how to light it, so they had to "harvest" from events in nature such as storms, and then keep and reproduce it for their domestic use. The first generation methods seem to have been to rub a stick or rope against dry wood or hitting two stones. Fire was discovered, but its use was designed. We carried it with torchlights, one of the oldest gadgets in the human tool belt.

Fire management allowed for our development. It sheltered us from the cold and defended us against predatory beasts. Cooked foods helped to better absorb nutrients and improve our health. Today, at the time when we cook the least for ourselves, TV cooking shows grow unstoppable. Something continues to attract us from the stove. Even today, Australian Aborigines baptize their children with fire. They smoke them at birth. Maybe it reminds us of when we became human.

Literally, fire made us human. It freed us from chewing endless chunks of raw meat and vegetable roots. Compared to their ancestors, the Homo species developed a weaker jaw, smaller teeth, and a much larger brain. They could have night activity for the first time thanks to the pyric light. Along with the domain of fire, pottery was born, maybe the first of the human arts. Fragments of fired clay 1.42 million years old have been found in Africa. Humans were working with clay before God himself created them from it.

Sapiens' first oral traditions emerged around a bonfire. We believe that the opposable thumb and a highly developed brain are differentiating factors that allowed sapiens to reign. But *Homo neanderthalensis* had a larger brain volume and corpulence. The evolutionary relationship between the two species is not clear; still, it is certain that *sapiens* prevailed for their more developed cerebellum and superior creative capacity. In particular, their ability to tell stories. Man-made fictions, from myths to economic conventions, have

The birth of tragedy

distinguished our species. Other animals can communicate or lie, but none narrates the way we do. We will see throughout the book to what extent we distort reality to dramatize a pleasant story. We have always been fascinated by stories. Thanks to them, we have been able to weave lasting bonds through generations. We create fictional herds along with millions of strangers, like those who belong to our country or religion. We do not find this striking trait in any other species in the animal kingdom. The value is not in the fire, but in our capacity to invent.

About 10,000 years ago, the Ice Age ended. Fertile land expanded. Humans went from hunting and gathering to settling and developing cultivation techniques. Thus appeared irrigation, plowing, livestock, and animal domestication for productive use. First civilizations arose in Mesopotamia and Egypt. The wheel appears around 3,500 BC.

We know that the wheel was not a sudden invention, resulting from eventual genius, but a continuous improvement of various components that developed over the years. It started with a tree trunk acting like a roller, on which the object to be transported was placed, then pushed until it reached another log. Or tied with ropes and slid on a roller track. Sleds facilitated the transport of the elements above the roller. They were positioned on the trunks, where slots were worked in the form of a rail in which skates were fit to improve stability. The wooden roller was still solid and heavy. The next step was to "sculpt" the wheels, thinning the structure of the trunk until obtaining a thin axis with solid wheels on which the rib was placed. The axle continued to belong to the same piece until the wheel appears as we know it: a rounded and independent part that fits on the axle, which today acts as the bearing of a car. The containers continue to be crimped until the trolleys appear. Wheels thin, making them hollow with wooden spokes, and so on.

Would we call two massive circumferences, undercut in a trunk axis, a wheel? Is a wagon wheel, with its hub and its wooden spokes, the same thing as a bicycle, with its rubber tires and metal rims? What was the innovation? Or were they all?

Figure 1: the evolution of the wheel. *Credit: Rudy Muhardika*.

Later the Greeks arrive. They bring philosophy, the mill, cartography. The atomic model is born in Democritus, although with little scientific rigor. The problem of irregular bodies, lacking the formula to calculate their volume, is solved by Archimedes by measuring the amount of displaced water, a technique still used today. We discover the density of bodies, a key development to civil engineering.

The Chinese invented paper, silk, the first instrument for making arithmetic calculations (the abacus), the compass, the first printing gadgets, centuries before Gutenberg. Also gunpowder, and with it all the beauty of fireworks and all the evil of firearms. Interestingly, the Chinese, Koreans, and Japanese adopted firearms late. The Portuguese brought arms to Japan in the mid-16th century. The island had been plunged since 1477 in an endless series of civil wars, a period known as *Sengoku*. However, gunpowder was not unknown to the Japanese, who had faced it almost three centuries earlier, during the two Mongol invasion attempts, in 1274 and 1281.

How is this possible? Some authors speak of a cultural factor. The Japanese perceived the sword as a symbol of their values: craftsmanship, honor, and respect; and firearms as the opposite, a shameful way to kill. It is also necessary to consider the different enemies that the Chinese and Europeans faced and another technology widely developed in Asia: equine dressage. In China, where the major threat came from the nomadic cavalry of the north and west, firearms offered few advantages. Archers were far more precise and deadly than arquebusiers and muskets, useless on horseback. So the Chinese invented firearms, but the Europeans perfected them. Today, Chinese, Japanese, and Koreans still have the most restrictive gun control laws and one of the lowest firearm murder rates in the world.[1]

The first planned cities are from Pakistan and India. However, the Romans improve territory planning and introduce aqueducts, the first major works of road infrastructure, and reinforced concrete. Also, the first newspaper-like publication called the Acta Diurna, published from 59 BC to 222 AD. These were handwritten political news-sheets on trials and military campaigns, which circulated daily.

With the fall of the Roman Empire, feudalism and the Middle Ages arrive. Crusades and the blast furnaces to melt metals. As Europeans sink into the shadows, on the other side of the Mediterranean and down to the Iberian Peninsula, knowledge flourishes among Muslim Arabs.

In the year 825, al-Khwārizmī wrote a treatise on arithmetic translated into Latin as *Algoritmi de Numero Indorum*. From here we get the word algorithm. This book describes the Indo-Arabic numbers, named after their Indian origin and later development in the Maghreb area. These are the numbers we use today in almost the entire world, including places without a Latin alphabet, such as Russia, China, or Japan. However, the symbols are not the interesting thing, but also the system. It is a base 10 numbering system with positional notation. The Greeks, Egyptians, and Hebrews already had their own, which comprised agglomerations of adding signs. It was necessary to invent a new symbol each time a certain value was reached. The Romans introduced a novelty: figures placed to the left of a larger one, subtract. To the right, they add up. The numerals represent multiples of five and there are only seven: I (1), V (5), X (10), L (50), C (100), D (500), and M (1000). They calculated the rest. For example, the numeral IV does not symbolize *per se* the number 4. We deduce it by subtracting 1 from 5. Vice versa, we calculate the number 6 by adding 1 to 5, since the minor numeral is on the right: VI. But this system has a problem. How to write 1776 in Roman numbers? MDCCLXXVI. We need nine symbols or numerals for such a small number. Europe did not introduce decimal positional notation until the 13th century, thanks to the Italian mathematician Fibonacci. Full adoption, replacing Roman numerals, did not come until two hundred years later.

Various civilizations had independently developed positional notation systems, including the Babylonians, the Chinese, and the Aztecs. The latter used a base 20 system. Why base 10 and why base 20? The simplest explanation is before our eyes: the fingers with which we count when we are children. Aztecs also considered those of the feet. The Mayans conceived the number zero in parallel with the Indians during the first years after Christ. Without the zero, other cultures with positional notation had to leave a blank when writing numbers like 109, writing it down as "1 9" and giving rise to obvious confusion with 19. The Babylonians instead used a system that may seem counterintuitive, base 60. However, it is the one we use to measure time: 60 seconds is one minute, 60 minutes is one hour.

Computers do not have fingers but also use symbols according to what they can understand. We talk to them in a base 2 system, binary, with only two symbols: zero and one. The method consists of electrical impulses. Zero is nothing more than a message: "There has been no electrical impulse." One, the opposite: "I have noticed a tickling." We call each of these numeral symbols a "bit." But how do computers know if there has not been an impulse, or it just has not come yet? It is necessary to set a pace. Therefore, all microprocessors have an internal clock, with a frequency measured in hertz. In the past, we could observe the speed of the clock on an external indicator of desktop computers. When a processor runs at 4 GHz, it means that that chip is "listening" for 4,000,000,000 cycles per second, four billion zeros or ones. For pragmatism, programmers stack them in base 8 or base 16 notations, as we will see later when we talk about IP addresses.

Al-Khwārizmī also introduced algebra in his *Compendium of Calculus by Reintegration and Comparison*, where he solves first and second-degree equations and tackles arithmetic. The Muslim influence is so great that many of the later advances in Europe during the Renaissance, such as Copernicus's heliocentric model, draw from their studies.[2]

The Renaissance meant the recovery of Europe through art and the return to Greek and Roman cultures. Many of today's Renaissance buildings in Italy could pass as original Greek or Roman buildings. Art arose from free time and abundant money. The bustling commerce of the Italian cities provided for it: Milan had arms, Florence cloth, and Venice and Genoa were neuralgic connections with Asia by land and with the Turks by sea. As a result, the first industrial agglomerations arrived. Incipient concentrations of capital in the hands of merchants arose along with the printing press, the microscope, the telescope, and Leonardo.

Commerce and industry grew voraciously during the Renaissance. Despite this, before the industrial revolutions, 80% of the population still took care of agriculture and livestock to feed themselves and the remaining 20% of the oligarchy. Today, less than 2% of the population in developed countries is engaged in the primary sector.[3] The productive forms were extremely manual. It was not until the beginning of the 18th century that this began to change.

INDUSTRIAL REVOLUTIONS

All revolutions have been characterized by a novel energy source, an original production process, and an invention. The First Industrial Revolution occurs around the middle of the 18th century with the mechanization and discovery of the coal-powered steam engine. It was born in Great Britain, which enjoyed a relative and prolonged peace, then spread to the rest of Europe. The textile industry is the first affected. Spinning and knitting machines appear. Production explodes using fewer workers. Profound socioeconomic and cultural changes also occur: These years witnessed the birth of the bourgeois class and the exodus of the rural population to the cities, giving birth to a new working class grouped in suburbs surrounding the factories. The population density was until then homogeneous, and the cities were few, small and underdeveloped. From the careful observation of their living and working conditions in the English factories, Karl Marx wrote Das Kapital, published in 1867.

The Second Industrial Revolution begins in the middle of the 19th century and ends with the outbreak of the First World War. It is characterized by the introduction of electricity, gas, and oil as powerful sources of energy over coal. Also by the substitution of iron for steel and the Henry Ford assembly line. This system was organized taking into account Taylor's model, which introduces the concept of scientific organization of work. Taylorism reduced all improvisation in production models, and the number of decisions a worker could make in the pursuit of productivity.

We can summarize it in four fundamental principles: the scientific study of work, that is, precise measurement of each production process; adjustment of methods according to the conclusions; systematic selection and training of the worker according to his aptitudes; and collaboration between employers and workers.

Taylorism works, the advances are amazing, but it has high human costs. The first concerns about technological unemployment emerge and survive to this day—will see how innovation and social rights are two related faces of the same coin. The labor movement, which had appeared with the last blows of the First Industrial Revolution, intensifies. The first modern general strike occurs in England in 1842. It is preceded by others at the local level, such as the one that occurred in Philadelphia in 1835, which ends with reducing the working day to ten hours. In Belgium, four occur in seven years: 1886, 1887, 1891, and 1893. In Spain, the one called by the unions in 1909 leads to the "Tragic Week," 78 people die. Argentina would also have its "Tragic Week" ten years later with 700 casualties.

Taylor published his work in 1911. The protest strikes intensified in the United States between 1912 and 1913. Lenin sums up the impact of his work in a phrase: "The most discussed topic in Europe today, and to a certain extent in Russia too, it is the system of the American engineer Frederick Taylor."[4] Faced with the unionist advance, Taylor defends himself: "There is a false belief that the increase in production leads to unemployment. It is poor administration systems that force the worker to limit his production. For, when he increases his work rate, the boss manages not to increase his salary. There are disastrous working methods that waste the efforts of the workers. Employers change procedures, according to the second principle. If each employee's compensation were related to their productivity, their performance would increase significantly." Taylor says this, but nobody listens to him.

The 20th century goes on with extraordinary social and industrial advances. But the same misgivings of yesteryear are still alive today, renewed under the concept of "computerized Taylorism,"[5] and criticism of globalization and neoliberalism. Today's knowledge society is organized under the same scientific prism that Taylor applied to the pre-industrial artisan society. Non-automatable tasks must be submitted to the same measurement, coding and mechanization process.

Finally, the Third Industrial Revolution brings us nuclear energy, electronics, Moore's Law, the accelerated development of microprocessors, and the first mechanical robotics, which facilitate factory manufacturing. For many authors, here we stay. For example, Jeremy Rifkin, who combines the Third with the current era, characterized by digitization. Others insist that we live in a Fourth Industrial Revolution, thanks to the unstoppable progress of wireless telecommunications technologies, the Internet, and artificial intelligence. (Perhaps an alternative energy source is lacking here. Maybe renewables will soon take on this role: solar, hydro, wind, biofuels, hydrogen, which have not yet replaced fossils.) It is a more common term in Europe, introduced by Klaus Schwab at the 2016 World Economic Forum.

Be that as it may, we are undergoing profound changes in manufacturing production modes and, the same as in previous revolutions, also social and political transformations that need to be understood, addressed, and legislated.

Before closing the section, let us pause for a moment to explain and analyze the consequences of Moore's Law. For which we will begin not with his enunciator, Gordon Moore, but with Douglas Engelbart.

Douglas's story is amazing: inventor of the mouse, a forerunner in fields such as word processing and graphical user interfaces, a pioneer in the development of hyperlinks, several years before the birth of the Internet. As a philosopher of technology, Engelbart introduced many concepts exposed in this book decades ago—collective intelligence, agility, radical innovation—Engelbart's Law states that the intrinsic rate of human performance is exponential, a key concept to speak of the current technological society.

Engelbart had devoured the work of Vannevar Bush. In one of his articles, called "As We May Think," Bush described the concept of memex, a contraption that would allow access to all of humanity's knowledge. We are in 1945 and people would travel for days to consult a single document. A machine that allowed access to any information was utopian. But Engelbart, who was 20 years old and had not even finished his studies in electric engineering, kept reflecting and developed several of his inventions on the concept of memex.

Fifteen years later he gave a keynote address in Philadelphia. It started like this: "What if everything in this auditorium was ten times bigger?" No one answered. But we know the answer: We would die sunk by our weight. The reason is that the larger the size, the volume does not scale linearly, but by a power of three. Consider King Kong, a character not only fictional but physically unfeasible, too. Although we can appreciate the gorilla, its reduced version, in freedom and captivity, King Kong is as impossible as the living dead or Count Dracula. The most modest, the original from 1933, measured six meters (almost twenty feet). In the 2017 version, Kong measured thirty-two meters or 105 feet. With an increase of one to two meters in height, we should expect about eight times its mass. Doubling the size does not mean doubling the volume; the volume takes into account not only size but also the width and surface area. However, the capacity of the bones and muscles depends on their cross-sectional area. Such an animal would need larger muscles and bones to keep itself upright. Therefore, we only find animals with thin support structures, such as spiders, in small sizes. If you were thinking of an ostrich, check the size of its bony hip first. The knees of a 32-meter-high gorilla would disintegrate under its weight.

"But the interesting thing," Engelbart continued, "would be to reason precisely the opposite: What would happen if we *reduced* ourselves ten times?" We would be ants with herculean force in relation to our size. We would jump tens of meters like fleas. Considering their size, insects are much faster than the fastest mammals. If it could have the body of a human one meter eighty-five, the common fly would move at about 1,300 kilometers an hour.[6]

Figure 2: Bush' article. *Credit: Dunkoman @ Flickr.*

The same concept—and here comes the subversive idea— could apply to microchips: On a micro-scale, properties change. We could create circuits with enormous computing power.

26 | The birth of tragedy

In that same room sat Gordon Moore, in his thirties at the time. He understood. That phenomenon is indeed what has happened with integrated circuits from the 1960s to today. Five years later, Gordon formulated Moore's Law. Three later, with Robert Noyce, he founded the Intel Corporation, the largest producer of microprocessors of the last century and still today. Moore's Law states that every two years the number of transistors in microprocessors[1] doubles. This means that, at the same size, the computing power is doubled, or we can maintain the same power by reducing the size in half. This has allowed for increasingly slim laptops and powerful and thin mobile phones. The costs of the chips have been reduced dramatically. The Pandora's box of innovation has been opened. Digital technology will become increasingly affordable. Its adoption, faster and faster.

BRIEF HISTORY OF CAPITAL NEEDS

Let us go back for a moment in time to talk about the largest company in history, measured by its market capitalization. It was founded in the 17th century and disappeared in 1799. Named the Dutch East India Company (VOC, in its Dutch acronym), it was, in fact, a concessionaire of the Netherlands, a country until a few years before under Spanish rule and still at war with them. Its mission was to exercise trade with Asia under a state monopoly regime. Flanders was an emerging region, and trade, particularly in the Indian Ocean, was gaining importance with naval development.

During the war, the Spanish took Antwerp, hitherto the capital of the new Dutch nation. The Duke of Parma gave its

---

[1] Moore's Law has slowed down in the 2010's and is likely to become saturated in the 2020's.

residents two years to abandon it. This marked the rise of a new economic power in Europe: the city of Amsterdam. It was there that from 1597 small companies began to form and to charter ships to the Dutch colonies in the East Indies, mainly Indonesia. Four ships in 1597, twenty-two in 1598, sixty-five in 1601. The VOC was a merger of six of these small companies by state imposition. In exchange, they were granted a monopoly on trade with the Asian islands.

Trading in the Indian Ocean seems easier on paper than it actually was. The Portuguese were present in the area and the English were expanding. In 1601, Holland was a union of seven provinces whose central government was in The Hague. A general council of 17 members governed the VOC, Heeren XVII, divided into six cities. But communications were so slow and the need for action so much that the VOC operated sovereignly while leveraging on the Dutch military power. In 1619, they conquered Jakarta and renamed it Batavia. There they installed the nerve center of their eastern trade. In the same year, the sister company, WIC, was founded to serve the West Indies, which covered virtually everything in Western India, and also included America. Over the next fifty years, the VOC and WIC unleashed a veritable world war against the English, Spanish, and Portuguese in Latin America, West and East Africa, the Persian Gulf, India, China, and the East Indies. By the end of the century, they had displaced the Portuguese from Asia and had settlements from Mauritius to Japan.

For the Dutch, investing in overseas trade was natural. The Netherland's financial power was reflected above all in the interest rate: While the Dutch borrowed at a 4% rate, the English did so at 10%. This allowed for a greater investment capacity. Besides founding the first stock exchange, they pioneered many other financial instruments: diversified portfolios, where citizens could invest small portions in several merchant ships, rather than all in one; or futures markets on the price of spices. We can still appreciate the fruit of that trade by walking along the canals of Amsterdam, where most of its quintessential houses were built during the 17th century.

The VOC was a super-company: It could mint currency on its own, declare war with the backing of the Flanders army, and colonize territories. It was also the first to publish its financial statements and was the forerunner of the first stock exchange, founded in 1620 to raise capital for merchant voyages on its ships. By the 1630s, its declared and inflation-adjusted fixed assets far outnumbered those of Amazon, Microsoft, Google, Apple, Facebook, and Alibaba combined. I have named these firms for their fame, although the largest organization by assets today is the Saudi Aramco, which manages the national oil of your country. Still today, part of the largest companies are oil companies. Similarly, many of the companies that resembled the VOC in size also did so in business activity, such as the Mississippi Co. or the South Sea Co. First, it was overseas trade, then, during the 19th century, banks, mines, and finally the oil companies. There, the biggest companies and also the biggest fortunes were made.

Overseas trade, banks, mines, oil companies ... there is something in common with all of them. Their founders needed enormous amounts of capital to start. This has been the case throughout history, from feudal proto-capitalism to the present day. All the companies mentioned—and in general those with capital-intensive physical assets—need colossal amounts of money to exist and subsist. Who can deploy fiber or build highways in a country? How many people can finance the works of a cogeneration plant? But something is changing, little by little. Something that, in recent years, has allowed what was unimaginable for the generation of our grandparents. Let us just look at the new stars in the constellation: software companies created by a few subjects, with few tools beyond a computer and some capital.

There is no need to stare too far back. In 2006, the six most valuable names by market capitalization were Exxon Mobile, General Electric, Microsoft, Citigroup, BP, and Shell. That is three oil companies, a bank, a multifaceted conglomerate diversified throughout many decades—an endangered species —, and a single software firm, Microsoft. Ten years later, the same ranking was taken by Apple, Google, Microsoft, Amazon, Exxon, and Facebook. Two of the six organizations remained the same. Only a single oil company, the rest concerning technology. Apple, the oldest, was started with a capital equivalent of about $350,000 in 2020 and a Mike Markulla loan of about $800,000. The seed capital to found Amazon in 1994 came from Jeff Bezos's parents. Google was born out of a Ph.D. project and attempted to sell for as little as $750,000 in 1999. Zuckerberg, Moskovitz, Chris Hughes, and Eduardo Saverin started Facebook with no seed capital in February 2004, although with a valuable network of students from Harvard, Stanford, Yale, and Columbia universities. Just a semester later, they received half a million from Clarium Capital.

Current technological advances and achievements allow individuals without financial backing, or at least with much lower capital needs than in previous centuries, to start businesses and generate huge incomes without fixed assets. There is the famous example of Uber, a transport company that does not own any cars; Facebook, a leader in content, but with no content generators on its payroll; or Airbnb, the largest accommodation hub that does not own hotels or buildings. These companies use a horde of online customers to write, design, transport, host, or build. One more fact, in case of doubt: The average lifespan of the largest companies has seriously dropped. In 1958, it was 61 years for US S&P 500 index companies. Today it is around 15 years.

What happened?

DIGITIZATION

Digital natives are subjects who have always lived surrounded by technology. People conceived from 1981 to 1996 earn the label *millennial*. Except in the case of a lack of socioeconomic support, all *millennials* have grown up on personal computers. For them, typing is as natural or more so than handwriting. And the difference feels stronger between those who have used the Internet since their infancy, born in the 90s and later. Nowadays children are raised with tablets. They start using them when they haven't even learned to speak. Meanwhile, those belonging to two or three decades before had to go through a period of adaptation to computer technology.

The same applies to organizations. We have them coming from a digital environment, and we find those who are already the age of our grandparents, type using only the index fingers and squint at the screen, without fully understanding each time an error message appears. The term *digital transformation* captures well the need for these businesses: a certain metamorphosis, a change between a previous state and a future one. Hence, *transformation*. Also, the participation of technology, hence *digital*. Adaptation is not enough, they need to mutate from the depths. For an adult, changing their habits is difficult, but most times it is enough to acclimatize partially. With companies, reform to the bottom is inevitable.

Today's children, those digital natives, have serious difficulties understanding how a gramophone or a music chain that reads cassettes works. But until well into the 21st century, we still preserved most of the world's information in an analog format: tapes, vinyl, videocassettes. What is the difference? In analog technology, we record the wave in its original form. On a recorder, we take the signal directly from the microphone and place it on tape. That recording is read, amplified, and sent to a speaker to produce the sound. This is what happens on a gramophone.

In contrast, in digital technology, we sample the analog wave at some interval and then convert to digits—zero and ones, hence digital. Then we store these bits on the digital device. On a compact disk, the sampling rate is 44,100 samples per second or 44,100 Hz. In other words: 44,100 pieces of numbers or bits stored per second of music live on a CD. Typically, they are 16-bit chunks, enough to represent the wave amplitude. And, in stereo recordings, one sample is stored for the left speaker and another for the right. That results in 1,411,200 bits or 176 kilobytes per second.[2] When listening, the numbers become a discrete voltage wave that approaches the original wave (see figure 3).

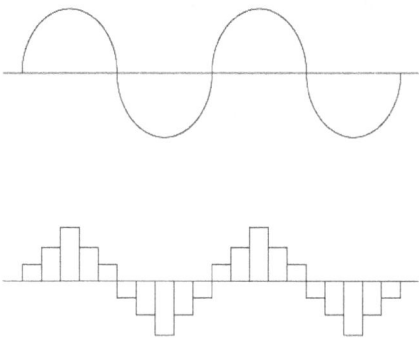

Figure 3: An analog signal and its digital equivalent. The width of the columns represents the sampling frequency. *Credit: Techno - Knowhow*.

As the discerning reader will have noticed, the digital conversion supposes quality loss, since we represent with fewer numbers and precision what the analog wave is expressing. When we digitize, we convert a continuous wave into small bits of a non-continuous representation—so tiny

---

[2] Per historical convention, one byte equals 8 bits. But early computers also worked with 4, 6 or 9 bit bytes.

that we find 44,100 in a single second. The precision of today's technology is so great that there is no difference between an analog and a digital recording, at least not a difference discernible by the human ear. Despite this, nostalgics will always prefer vinyl.

Digital signals support something else: compression. Digitization itself is compression, converting a continuous signal into portions. But there are also algorithms capable of detecting patterns in the beam of zeros and ones that make up the signal and reduce the number of digits necessary to represent the same, or something very similar, depending on the level of compression.

The most famous compression algorithm of all time is perhaps MP3, developed by Fraunhofer in Germany in the late 1980s and licensed to Technicolor. Its patent expired in 2017. We said that music is sampled on a CD at a rate of 176 kilobytes per second. For a 3-minute song, that is 32 megabytes. If you are young, I guarantee that in the 80s and 90s, this was a lot. The first edition of Monkey Island, a video game that marked childhoods, was only 16 megabytes. Sending a song of that size in the early stages of the Internet was utopian. But with MP3, we compress by factors of 10 to 14, the song is reduced from 32 to 3 megabytes or fewer. How is it possible? By means of a trick called noise shaping, which eliminates unnecessary frequencies. There are certain sounds that the ear cannot capture, or that are perceived less than others. When two sounds are produced at the same time, we tend to hear the loudest and only that one. If there is a snare on the drums, we can remove some notes from the guitar. MP3 does just that: subtly remove portions of a song, without us noticing. What we are really doing is deleting, not compressing. The more we delete, the more we will notice, but the compressed file will take less space. We configure this with the bitrate, the number of bits per second that we encode. At a 40 kilobytes rate, it becomes impossible for the human ear to differentiate the quality of that compression from its original signal. That is to say: In the worst case, the algorithm can divide by 4—from 176 kilobytes to 40—the size of a song without us knowing.

This *modus operandi*—erasing imperceptible parts—is shared by all image and sound compression algorithms, such as JPG or MPEG. In contrast, file compression algorithms, for example, ZIP, work differently. We cannot delete anything, so as a result, they try to represent the information in a summarized way—but ways we can reconstruct accurately. For example, for a stream of bits: 11000000011110000001, you could invent an algorithm that represents it by the bit followed by its number of repetitions: 1207140611—read "one twice, zero seven times, one four times," and so on. In this example, I can represent 20 bits with 10 numbers.

Computers originally operated on analog tapes and only incorporated digital transistors from the 1960s. Since the end of that decade, digital storage mechanisms such as the LaserDisc already existed. But it was not until the late 1980s, with the compact disc, and especially since the 1990s, with the appearance of the most famous compression formats—the aforementioned MP3, the JPG image algorithm and the MPEG video algorithm—when we can say that we have fully entered the digital age. And it is not even until approximately the year 2002 when it is calculated that the amount of information stored on digital media exceeded that preserved in analog formats.

It may seem contradictory, but still today organizations keep immense amounts of information in non-digital format, mainly paper. And other data in a digital low-quality format, not taking advantage of the possibilities offered by the latest technologies. These are the so-called legacy systems: ancient and inflexible software, information stored on local hard drives that are not accessible and prone to failure, databases in Excel, and so on. Even when it comes to bits, not all bits are equal. A bit in the cloud, accessible by different profiles under different security conditions and from various devices, with layers that allow its analysis and protection, is preferable to the data in local storage or paper. We will see all this in greater detail throughout the book.

Everyone adapts the term digital transformation to their reality. For a digital marketing company, it will not be the same as for a company that sells hardware infrastructure. But they can all be right. Digital transformation covers several layers of the value chain, if not all. For consumers, "digital" is anything that comes from the Internet, like apps, or electronic devices like mobile phones or smartwatches. For companies, "digital" includes information systems' infrastructure, communication processes, human resources, supply chain management, and many others. For a CEO, it means that we are making updates to our e-commerce so we can tell our investors we are keeping up to date. For a CMO, we are spending too little on digital marketing. We are spending way too much, according to a CFO. And for a CIO, we are buying a new customer management system, and for human resources we are automating, downsizing, and moving work abroad to save money.

Conversion to digital is necessary, but the resistance is titanic. It's a shame. Innovating has risks, but avoiding it has its counterpart. Most organizations are slow to react and thus stagnate. An actual transformation breaks down the barriers of hierarchical inertia and turns the organization into an adaptable animal. Increase competitiveness, revenue, market share, reduce customer acquisition and operating costs. It attacks key elements of the organization that must be addressed in the business model: How do we generate revenue, gain customers? What products do they want, what

What we talk about when we talk about innovation | 35

internal processes do they need to change, new forms of work, decision-making, analysis?

Renew or die. In the words of Douglas Engelbart:

> *Digital technology will become increasingly miniaturized and affordable, and its injection into all levels of business and society will become increasingly widespread and rapid. This will cause a disruptive ripple effect in society like never before seen, shifting us onto an unsustainable trajectory where important challenges become increasingly complex and urgent with potentially disastrous consequences to humanity if this phenomenon is not well understood and adequately addressed. The vast majority of our organizations and institutions, which steer the boat we are all on, are severely underestimating the magnitude and speed of the curve, and thus are aiming too low and operating too slow. It is no longer an option to get incrementally smarter and faster. Organizations must become exponentially more intelligent and agile, using successive gains in Collective IQ to accelerate progress toward that goal. Those who lag will be rendered increasingly ineffective.*

THREE NEW TRENDS

How much does it cost to use Facebook? How much to tweet? Searching on Google means we have to pay? And if so, in what currency? How is it possible that there are companies that are worth so much and still it costs us nothing to use what they offer?

We produce two-and-a-half exabytes of information every day in the world.[7] It is a lot of information—specifically, 2,500,000,000,000,000 bytes are created every day. If Facebook was a country, with its unique monthly users, it would be the most populous on Earth, far above China and India.[8] In fact, Facebook users are around the populations of these two countries combined. If we made a human chain every time we post a tweet, holding hands, we would go to the moon and return every day. Each individual on this planet, including the elderly and children, receives an average of 40 emails a day.[9] When we watch spy movies, the intelligence of the special agencies of the British, Russian, or American governments amazes us. But Uber receives 14 million trips a day and knows who makes them, from where and to where. Google records all our movements, searchable on Google Timeline and preserved in its databases. If it weren't for Moore's Law, our hard drives would have to become the Colossi of Rhodes to withstand all of this. And we wouldn't even make it that way, due to lack of space: IBM put the first hard drive with a gigabyte capacity on the market in 1980. It weighed 250 kg and measured like a refrigerator. If we had to use that technological wonder today, at our current information production rate, we would cover the entire surface of Spain with hard drives in half a year and all of Europe in a decade.

Almost all organizations generate staggering amounts of data. Some do not know it and do not capture them. Others, however, have been able to offer free products thanks to the value that the businesses derived from the data generated for them. Those with physical assets, such as hotel chains, telecommunications companies, or infrastructure concessionaires, get information on those physical assets. The engineering companies carry out field studies from which they could draw more conclusions than necessary for their project. Even companies that do not have this arsenal of assets often have information systems or a valuable customer database. Products also generate information. Amazon not only sells physical items. It also knows, through its website, the preferences and concerns of consumers, and issues recommendations. Sports equipment companies like FitBit sell gadgets that allow us to monitor our heart rate in a race, but they also accumulate a lot of information about how much, where, and at what time people run.

Moore's law and the lowering of the cost of chips have enabled three phenomena that make this possible:

What we talk about when we talk about innovation

1) The **platform business models** that organizations without large fixed assets operate, putting into agreement subjects that offer and demand services. Such are the cases of Uber or Airbnb. These allow generating businesses with lower capital needs than the entrepreneurs or governments of several centuries ago had to face. Before it was the Dutch government monopolizing an activity, now there are four university classmates.

2) The increasing **connectivity** has confronted two vast groups on the mat: On the one hand, enthusiasts eager to collaborate on an interesting project, even for nothing. And, on the obverse, a growing clientele that increases the value of these platforms; the more people use it, the more it serves us. The more it serves us, the more people are interested in collaborating. It is a virtuous circle. It is because of this new connected crowd that platform models are made possible. We call this effect is "network externalities." In economics, externalities are the events or costs caused by someone who did not intend in principle to cause them and which are not considered in their price. For example, when we buy a car, we do not pay for noise pollution—and the emission of gases is not assessed in most of the world either. Externalities are usually negative. However, if we refer to an electric car, things change: The more there are, the more chances we will have of developing a network of charging points, which is insufficient today. In the same way, a smart car generates data. These data are prone to be processed and studied to improve their autonomous behavior. In the digital world, network externalities are generally positive.

3) The emergence of analytical models, big data and the **growing importance of data** on other merchandise in

today's world, which allows certain organizations to give away their product for receiving information.

Who generates such valuable data? The crowd connected through platforms. The effects of these three trends feed into each other: Low-capital platforms offer free or low-cost services. This makes it easier for people to use it and increase their value or complementing the product. Finally, this number of users generates formidable amounts of data prone to be mined and applied to analytical or predictive models, which makes them attractive candidates to adopt artificial intelligence.

Figure 4: Cover of The Economist on May 6th, 2017. *Credit: The Economist*.

The circle is closing: What has made it possible for companies to give away for free demanded products is that the data has become a much more valuable asset than almost any other product if we know how to mine and exploit it. The cover of *The Economist* on the 6th of May, 2017, defined it this way: "the most valuable resource in the world." Companies that understood this have thrived and smart platform owners manage the supply and demand encounter with excellence.

WE ARE NOT ALONE

2

"Loneliness is fine. But you need someone to tell you that loneliness is fine."
—Honoré de Balzac, 1799–1850

Platform models capitalize on the economy instantly and for free. How? They put the demand in touch with its counterpart, the supply, in exchange for a commission of some kind, *which may not always be money*. This centuries' markets are the new evolution, created from software but with identical essence. Like regular markets, they are more valuable as more people join in its use—WhatsApp or YouTube both become more useful when more people participate. This new economy, which has been with us for some years now, continues to take entire industries by surprise today.

In the past, music lovers needed to go to the store during opening hours to listen to the latest LP of their favorite band. They also needed enough money from a decent salary or weekly pay from their parents. Another option would have been waiting until the LP was broadcasted on the radio, or to meet the right partner to copy it. But at some point, P2P[3] applications, like Napster, came along and changed the entire process of purchasing music. They were also platforms that brought together users who did not know each other, to exchange music. The music industry vilified and banned these programs under the threat that music would disappear. Of course, music did not go away. On the contrary, people listen to music more than ever.[1]

From a music consumption standpoint, Napster and other pioneers are direct ancestors of iTunes or Spotify. The latter just converted the new consumer format to its current paid-for model, monetizing it under a monthly subscription or pay-per-view. In contrast, YouTube, where the same producers as VEVO—a joint venture of Universal, Sony, and EMI—upload their video clips for consumption, is free. YouTube's business model is different: it makes money from ads and not from consumers—except for those who voluntarily subscribe to its premium service.

Together, these programs changed the nature of selling recorded music, packaged and locked up in physical format. It was inevitable: digital copies are much cheaper and allow for unbundled sales. The music industry is far from disappearing and is, in fact, growing further and further.

There are two viable approaches to creating a platform. The first is to generate a new market that intermediates between supply and demand, with low commissions—or for free. In a new market, demand increases as the cost decreases; a market made of software, for example, is cheaper than opening a store

---

[3] P2P, or peer-to-peer networks, allow the direct exchange of information between interconnected computers. It is the architecture under the famous file sharing applications, mainly music and movies, which appeared at the beginning of the century, such as Napster or eMule. In reality, the P2P architecture is respected by many other services, such as Skype calls or bitcoin transactions.

—and getting cheaper, thanks mainly to Moore's Law—, and the ease of online access would increase demand.

The second approach is to turn a product into a platform by embracing the notion of add-ons. Disguised and in the background, apps captain the demand for the iPhone. They were, at heart, the flagship product. The touch screen was not entirely new, nor were the operating system phones—the first pads already had their own applications. In a sense, the iPhone was more of a software revolution than a hardware revolution. It changed the concept of the phone, not because it had a better keyboard or a brighter screen, but because we started using it differently. After Steve Jobs gave the green light for others to develop apps for his phone, iPhone sales skyrocketed. It wasn't just the phone alone, with its touchscreen, its screen scroll to find contacts, its eye-catching design, or even the Apple logo on the back. There were also hundreds of applications that allowed you to follow sports scores, get the weather forecast, or listen to the radio. Something magical happened: Apple was able to turn its physical product into a platform. It no longer attracted just people who wanted a phone—with its many apps, it also appealed to those interested in measuring their pulse, listening to foreign radio, reading the news on the subway, or checking their bank balance on the street. This change meant an additional and unexpected margin—of course, not all applications are free, and Apple collects 30% of these sales. In the process, they also collect data on behaviors and preferences, which is a sweet deal for Apple. Moreover, Apple decides who can access its platform—a necessary and valuable step as it allows control of both development and quality retention of the "product-platform."

Let us stop and reflect for a moment on the internet-connected crowd and the rise of remote work. Linux operating system distributions (such as Ubuntu or Fedora) were developed from a common open-source kernel—the coding core—created by Linus Torvalds in the early 1990s. Linus is still responsible for this kernel, but entire companies have been built upon it. An example is the company Red Hat, which today has more than twelve thousand employees and writes layers of software that complete and complement that kernel. Projects like Wikipedia were only possible thanks to the participation of thousands of people who populated its articles. Jimmy Wales and Larry Sanger devised the internet platform, but the value of their encyclopedia comes from the information provided by other people—the add-ons. With widespread access to the internet, the connected multitude has eclipsed the ancient central repositories of knowledge. The differences between the web and the world's libraries illustrate how different the two are; the latter holds millions of books, while the former holds music, images, podcasts, videos, or virtual reality. The internet is democratic: no one is in charge of all the content on the web, as long as governments don't have the power to break its neutrality.

This democratization, along with the concentration of many elements into one—the telephone or connected computer—has contributed to the emergence of this new mass economy. More and more people have access to the internet, and it is becoming easier to do things that were previously only reserved for technicians, such as creating an online shop or writing an app. Wikipedia is an excellent example of why crowds can generate valid and compelling content. All previous encyclopedias operated similarly: they had a team of experts who drafted the articles and were paid by the publishing organization in charge of creating these collections. The number of copies printed was limited by the sales they made and their time was limited to a particular number of articles. Encyclopedias were rarely updated; once a year at most, one could receive an additional volume with corrections and updates. Today, on Wikipedia's website, corresponding articles display how an event has been updated within seconds of its occurrence, be it a political election or the end of a football match. Attempts to sabotage information do exist, but they are corrected in a few minutes. Controversial issues that need revision are padlocked and discussed behind the scenes before any edits are made.

What would have happened if no one had been interested in writing articles for Wikipedia? It would have failed, of course. This phenomenon is known as the "ghost town problem." No matter how high quality a platform is, no matter how magnificent and noble the idea behind it, the success of the product depends on the influx of the public. For example, Google, with its social network Google+, was unable to compete with Facebook, which had already gained the attention of the public. There are some ways of solving the ghost town problem, including:

- **Attracting new demand**, as TripAdvisor or Yelp did. Yelp is a social network for evaluating businesses of all kinds and is fairly popular in the United States. TripAdvisor is used with similar frequency in the rest of the world and has a similar model. Both TripAdvisor and Yelp started by creating a business directory and a sector-specific vertical search engine for restaurants, hotels, and general tourism. First, they had some information gathered by them, with which they reached the customers. When the businesses started to feel that they were garnering traffic from these websites, TripAdvisor and Yelp started charging them with a cost-per-click model: every time a person clicks and books through the platform, the businesses are charged a commission. Voilà: we already have businesses and consumers participating.

- **Attracting new supply**, as Clip in Mexico or OpenTable in the United States did. In both cases, they had an exciting product: a payment system in the case of Clip, and a piece of software for internally managing orders and reservations in the case of OpenTable. OpenTable went from being an internal software for restaurants and bars to an online system that allows customers to book directly. Clip has not yet taken the same steps as OpenTable, but its co-founder is clear about their plans to progress in a similar fashion: "in a period of four to five years, four out of ten shops will carry out their transactions with Clip. This is the opportunity to turn Clip into a trading platform for people and businesses."[2]
- **Help on the weaker side**. In dating apps, women tend to be more reticent. That is why some apps like Bumble provide them with certain advantages while using the app. Bumble allows women to choose whether to tilt a conversation after a match. This gives them a second chance to think about it and makes them feel more comfortable using the app.

Opening a platform has its risks. You do not know the developers of your initial product. You do not know what will happen. That is why a safer alternative has been staffing on demand, leveraged on the so-called *gig economy*. Specialized agencies—which in turn serve as platforms—offer open markets where companies and individuals meet and agree on projects. These same markets that create the platforms are open to on-demand hiring. For example, if you are thinking of setting up your own business and you need a logo or an announcer to record your advertisement, Fiverr could be a good option. If you require a project more complicated like software development, engineering, or customer service, UpWork might work. Large companies that need something much more customized, however, would use Gigwalk or a similar agency. And of course, in the Spanish-speaking world, the same concepts are starting to emerge, like Workana or Freelancer. Now, there are apps to walk your dog, to send food to your home, or to find a lawyer.

The fact is that phenomena similar to Apple's need not be reserved exclusively for huge organizations; you can leverage this new reality from your home. The future workforce will no longer be a group of employees arriving at the office or factory every shift and leaving after a certain number of hours. It will increasingly be composed of temporary or project employees who will work on a variable basis, paid by the hour, and in many cases operating remotely. Today, more than a third of all workers in the United States are employed on a project contract basis. In Canada, it is more common to talk about an hourly rate than an annual salary when negotiating fees. The uberization of the economy will be a fact, which is why political and legal adaptation is so important to avoid precarity—especially in non-specialized sectors. Most authors seem to think only of software developers, but this change will involve many more fields.

CONNECTIVITY

Humans are social animals. We belong to the herd and need to feel part of something more significant and meaningful—this is what nationalism feeds on. The need for affiliation ranks third in Maslow's pyramid, above basic physiological and safety requirements, and below demands for recognition and self-realization, both of which are also intimately linked to the group to which we belong. Credit is always in relation to a group: our perceived success, the trust we have in others, and that others have in us. Finally, self-realization is only possible in a group context and relates morally to the customs exercised in a specific time and social context.

One of the pillars of the current industrial revolution is the significant development of communications technologies since the 1980s. Ubiquity is the mark and destiny of many extreme changes that we are witnessing today. Of course, this does not imply existing in all places at one time, but suggests the ability to interact with everyone from anywhere. It refers to the integration of multiple objects and their functionality into a single mobile device, and the connectivity that allows that device to interact with other communications systems—beyond the calls and store information in the cloud.

Let us think about how humans worked in the '50s and '60s. It is not a distant time: cars with explosion engines were already relatively democratized. We had the Beatles and Chuck Berry, we saw mini-skirts. We communicated by mail or dial-up phone and the news was in newspapers once a day or on television with regular frequency. Documents were mostly typed, so an office's pace was slower—which was perhaps a good thing for our hearts. Think back to the 1990s, when we used fax machines that allowed us to transfer photocopied documents reasonably quickly: about one page per minute. There were already personal computers and telephones in every workplace. We used calculators, lots of notebooks, pens, and paper calendars and maps. Music was first heard personally and privately with the Sony Walkman. Hesitantly, the first laptops appeared. Taxi drivers knew every street and every inch of the city, no matter how big it was. Those of us who weren't taxi drivers had to check routes in giant coil-bound books that held street maps. We looked for telephone numbers in the Yellow Pages. When we had a doubt, we consulted the encyclopedia for answers. Bets among friends were exciting, but they took days to resolve. We had no Google, no Wikipedia to consult. Some offices had analog alarm clocks to warn of important meetings, and all had notice boards made of cardboard or blackboard. We took photographs with Kodak film cameras until the invention of digital cameras, whose brilliance spanned only a decade.

In 1990, the Office suite arrived, and with it the indispensables Excel and PowerPoint, which replaced slides and transparencies that were placed on a beamer and written with alcohol markers. Online dictionaries and the first advertising websites appeared to replace paper magazines. Email and PDF would soon eliminate the fax and mail service, which was increasingly limited to packages. GPS devices with route maps would kill off paper maps and Michelin guides; mobile phones with built-in navigation apps then arrived and wiped out the TomTom. Calendars were digitized. Hard cardboard folders with sections labeled by colored tabs—ancestor to browser tabs—disappeared. Meetings were easier to organize. The desktop phone was replaced by the cell phone, then by Skype, then WhatsApp, Telegram, and others. We began reading online newspapers more and more, and the first social networks arrived—some specialized (such as LinkedIn) for job and talent searches. And the phenomenon continues.

In the 1990s still, children would go out into the streets and remain out of range for hours. They chatted on the intercom, and on the last day of school, they asked each other's addresses so that they could send each other letters. In many villages, telephones did not reach all houses, and people came to talk from call centers, forebears of the cybercafés, also practically extinct. The ones running late got lost and had to wander around the city in search of their friends, who had previously agreed to meet in a precise place and at a definite time with no need to confirm. Later, from home, they would phone each other using a coded language since there was only one phone in the whole house. Calling the boy or girl you liked meant confronting a guard who answered first. At music festivals, there were long lines to call from the booths just as there were lines to the bathrooms.

By the year 2000, there were about 400 million internet users in the world. Today there are about four billion of us. Not only is internet penetration increasing immeasurably, as are social network consumers and mobile phone customers. The ratio of data traffic to total mobile traffic is growing exponentially every year. This means that people are increasingly using their mobile phones to run Internet-connected applications and less for what a phone is supposed to do: make calls.

GPS chips cost thousands of dollars in the 1980s; today, they cost between two and three. Maybe by the time you read this, it will be less than two. Every phone today has a GPS that has practically no effect on its cost. The Internet will become an invisible commodity: we already take it for granted on the phone. The conversation has matured changing from a "connect and search the internet for tomorrow's weather" outlook to a "look at tomorrow's weather," one just like that. The connection is an invisible step that we take for granted; technology democratizes it before we can realize it.

THE CLOUD

Not so long ago, the Microsoft Office suite was bought in a cardboard box, filled with floppy disks or CDs — considerably bulkier than a hardcover novel. Microsoft has been able to adapt to new times and introduce different types of innovation. The Office suite is in itself a "product package" type of innovation: It went from marketing different types of software to creating a complete office suite and selling it as a whole. In the late 1990s, Microsoft also wanted to package its Windows system and its Explorer browser, which earned it a

What we talk about when we talk about innovation | 51

lawsuit for corrupting competition law.[3] In recent years, Microsoft opened its suite to third-party development, allowing the placement of add-ons on top of Word, developed by third parties. Finally, it published Office365, a suite that allows users to work wholly connected to the Cloud and through an Internet browser, without having to install any software on a local computer.

We all use cloud technologies daily, even if we do not realize it. Not many years ago, mail servers—under a protocol called POP3—downloaded emails to the local hard drive. This way, if the disk was corrupted, all the emails would be lost unless we had them saved somehow—for example, old versions of Outlook allowed to compress all our emails in a PST file. That changed with the IMAP protocol. Today we all use Gmail or other online mail services without realizing that our information is hosted on one of Google's servers, far away from our machine. If our computer breaks down, we can borrow another one and still have our mail there. Most of the information we produce today is in the Cloud; our files in storage services such as iCloud, DropBox or OneDrive; our photos on Instagram or Facebook; our messages on Twitter; our databases and spreadsheets in Google or Microsoft cloud suites—if we use Office 365—Slack, Salesforce, the list is endless. Our life has moved to the Cloud without us realizing it, yet many companies still use physical servers in their own data centers. Why?

Cloud technology, for storage and computing, is interesting for companies of all sizes for a variety of reasons. The main reason is undoubtedly the flexibility in service level and costs. If demand increases, it is much easier to extend the contract in the Cloud than to purchase additional hardware. It is usually possible to contract online or by phone to our account manager assigned by the service provider. Cloud-based services are ideal for businesses with increasing or fluctuating bandwidth demands. And vice versa, if you need to downscale again, flexibility is built into the service. This level of agility gives enterprises that use cloud computing a real advantage over their competitors. The same applies to the cost model. Cloud computing reduces the high cost—in money and time—of buying hardware. It pays for itself on the fly, which is a more cash-flow friendly model, not to mention that, for accounting purposes, companies can play with their CAPEX and OPEX.

Another notable advantage of cloud computing is that we outsource issues such as physical and logical security and software upgrades to the provider of the Cloud. The mental burden on our hardware architecture or the need to hire system administrators is considerably reduced, which is especially crucial for small businesses that lack the required cash and expertise. Providers take care of this burden by deploying regular software updates, including security updates. They also remotely wipe data from lost laptops, so it does not fall into the wrong hands.

Cloud computing also imposes discipline on companies to make their data available via APIs and to develop in-house applications rather than having all their data and calculations on paper or in the Cloud. An API is simply a "library" of function names that a piece of software is capable of performing. Without providing us with the source code, the API enables us to use that software by calling those functions. It is nothing more than the ability of different software to communicate without mixing code.

Let us look at two concepts frequently used interchangeably: microservices and web services. They refer to similar concepts. A microservice is a type of software architecture. In the past, software was written as a monolith, one single application that includes everything. Everything resided in the same place: the graphical interface, the business logic engine, and the data access layer. Microservices are a way to build software, breaking down into sections parts of code that are dedicated to fundamentally different things. Later on, we will talk about the model-view-controller pattern, which breaks the code into three pieces. Therefore, a microservice is nothing more than decentralizing the source code. What do you get out of it? Flexibility, limited responsibilities, less rework, and much more. This philosophy applies to any environment.

A web service is nothing more than the concept of a microservice applied to the communication between two machines in a network through the HTTP protocol, i.e., the internet. An API acts as an interface between two different applications so that they can communicate with each other. This means that an API is the catalog of everything the microservice has implemented.

If you think that you have never written software and never will and that these are just developer technicalities, let me tell you that you are right. Perhaps one detail is missing, however: the opportunities this concept opens up are immense for everyone because APIfication is everywhere and available to anyone who wants to make use of it.

For example, if we wanted to write a piece of software to know where our favorite team will play next year, we could generate a map by building on the Google Maps API, accessible and usable from cloud.google.com/maps-platform. The Maps API is a paid service, but simple situations, e.g., publishing your business' address on your website using maps, are free and do not require programming skills.[4] Not interested in doing it yourself? Remember the gig economy. You can hire someone in any employment platform to do it for you. If you are a beef wholesaler thinking about creating a system to locate your distributors, you should know that your developer will most likely do so using Google's APIs. And, of course, you can consult the costs of consuming it yourself.

We also find internal APIs. Steve Yegge narrates in his blog *Stevey's Blog Rants* the anecdote of the internal mail that Jeff Bezos sent to all Amazon employees in 2002, known as "the API mandate." He requested for the following:

1) All teams will henceforth expose their data and functionality through service interfaces.

2) Teams must communicate with each other through these interfaces.

3) There will be no other form of interprocess communication allowed: no direct linking, no direct reads of another team's data store, no shared-memory model, no back-doors whatsoever. The only communication allowed is via service interface calls over the network.

4) It doesn't matter what technology they use. HTTP, Corba, Pubsub, custom protocols — doesn't matter.

5) All service interfaces, without exception, must be designed from the ground up to be externalizable. That is to say, the team must plan and design to be able to expose the interface to developers in the outside world. No exceptions.

6) Anyone who doesn't do this will be fired.

7) Thank you, have a nice day!

Even if you do not want to get to this point, it is useful to get your mind used to think under a "microservices communication" concept for all company activities. When teams can access, edit, and share documents at any time, from anywhere, using online applications, they are more productive and can work faster. Using these shared documents reduces working in silos and promotes collaboration. Microservices communication also allows workers to update documents at any time and maintain full traceability of changes made. The more people collaborate on documents, the higher the need for document control that regularly only specialized software is capable of doing. This is greatly facilitated in cloud-based work environments: from office suites — such as Google's or

Microsoft's—to specialized document managers by industry. And of course, anyone can work from home or anywhere else with an internet connection.

INTERNET OF THINGS

How many engines does a car have? One? Indeed, there is only one combustion engine. Just as important, however, are the endless numbers of electric motors. That is why, if we run out of battery, the car will not even start. We have an engine for the windscreen and another for the windows, which are no longer cranked by hand like they used to be. The air conditioning system contains several motors on its own. We do not even notice them there anymore. Like the internet connection, they disappeared from our consciousness. Soon we will see a convergence between the computer and the industrial worlds, thanks to the collapse of prices of geolocation chips and the growing connectivity.

The first problem we encounter with the Internet of things is its name. It was proposed in 1999 by Kevin Ashton of MIT. An earlier term, coined by Mark Weiser, "ubiquitous computing," seems a better choice. We are not talking about a different internet with which furniture entertains itself on rainy Sundays. In fact, many of the objects we are referring to do not even connect directly to the Internet, but use other short-distance communication protocols, such as NFC, RFID, and the best known: Bluetooth.

The Internet is nothing more than a set of computers controlled directly by humans and connected under a standard set of protocols. Devices are uploaded to this ecosystem, intertwined under a single language. But let us not expect our fridge to start googling—although it might. They use those same protocols to communicate a series of parameters so that we can give them orders remotely—if they would accept orders. When we talk about the IoT, we usually refer to inanimate objects, but it also applies, for instance, to herd control. And it does not appear that the day is coming when the sheep will obey us by remote, electronic commands. Another example is to find out the driving style in a car, which is an inanimate object but controlled by a human. The purpose of measurement is the human being.

So, what are the things that we can connect? We generally divide IoT devices into six categories: personal appliances, home automation, cars, health care, industrial and business manufacturing, and smart cities. Although these categories exist, almost anything can be managed.

Today we already perform many remote actions using our mobile phone. We can, for instance, request a taxi using one of the mobile applications that allow us to do so. The act of asking in real time for a car to pick us up a specified location, with an estimate of the cost and a confident certainty that the person who will come to pick us up is someone nearby, is in a sense, a typical action of the Internet of things. We do not interact directly with the driver, but there is an intermediary: an app installed on his mobile phone.

Think of your car and how you always forget whether you locked it or not when you are already comfortably sitting on the couch at home. Although many already have an automatic locking system, triggered by the distance from the key, you now may have a way to know if it was indeed closed, or how much gas it has left. Imagine something simpler, like the lamp you left light in the room. It is an elementary item with two functions: switching on and off. Why not install a wirelessly connected actuator that allows you to switch it off from your phone? Or regulate the heating to keep the house warm for a few minutes before you arrive. Or think about your coffee maker, which you can leave charged at night and activate it when you wake up. Although it seems that you will never need a coffee maker connected to the Internet in your life, we still thought about mobile phones many years ago.

Therefore, IoT is a set of machines that perform actions under remote command, collect data, and communicate them to us. This communication is executed by uploading it to a network.

The term is abstract, but the technology is very tangible. It is already applied in many places to collect information about almost anything you want to control. We have even gone from exclusive human-machine communication to introducing intermediate machines that filter, automate, or work in some way with the raw information sent. This set of interactions is also called machine-to-machine, or simply M2M, to refer to communication between two remote machines. M2M is a general term because under the hood and in detail, IoT implementations use different connectivity models, each of which has its quirks. Four are distinguished: device-to-device, device-to-cloud, device-to-gateway, and back-end data-sharing. We will not go into detail, just notice there is a bunch of heterogeneity—and also flexibility—in the way devices are connected and work.

The number of applications for the IoT is unforeseeable. Some are very complex and others we use almost daily without realizing it, such as the point of sale (POS) terminals we use to pay by card in shops. A POS is nothing more than a card reader that contains a SIM card—some use WiFi—to be connected wirelessly.

In short, every M2M ecosystem has at least four components:
- the monitored *thing*;
- a sensor that captures data, like a shelf counter; or a device that performs a remote action, like crediting;

- a communications technology, which includes the network and its protocols, wired (PLC, Ethernet) or wireless (Bluetooth, WiFi, mobile network);
- a server that collects the information and typically communicates with some other core systems of the organization, such as an ERP.

These four elements give rise to an incessant number of different lines of research: types and networks of sensors, next-generation networks, ways of interconnecting heterogeneous devices (both physically and logically), cybersecurity, distributed systems, fault-tolerant architectures under extreme conditions, edge computing, and a very long etcetera.

The concept of using computers, sensors, and networks to monitor and control remote devices is not new and has existed for decades. Intelligent domotics became popular in the '80s. The first communication protocol imagined for it is from 1975, the X10. It worked through the electrical network. Its democratization, however, is the novelty, enabled by the connectivity, reduced chip cost and size (thanks to Moore's law), advances in data analysis, and cloud computing. The Internet of Things has fantastic implications for the way we live, transforming cities, homes, and any environment in which we live.

If the trends and projections for IoT growth become a reality, it will force us to change our way of thinking. In this new world, the most frequent interaction with the Internet will come not from humans, but connected objects. It will be a wholly joined planet. And, as with the Internet, we will forget it exists.

In fact, the number of connected IoT devices has already surpassed that of personal computers between 2013 and 2014, and mobile phones between 2017 and 2018[5]. We already see connected objects everywhere: in the production industry, with machinery that controls manufacturing processes, assembly robots or temperature sensors; in urban logistics control, managing traffic lights, bridges, train tracks, urban cameras; in meteorology, with information from atmospheric, meteorological and seismic sensors. There are also applications for transport or the energy industry. It will soon be democratized towards homes.

But there are three that I would like to discuss briefly: health care, smart cities, and connected vehicles.

HEALTH CARE

Universal education and health care is a matter of debate in much of the world. Public health systems have been harassed in recent years under an alleged argument of unsustainability. They are complemented in this fight by the retirement system and an aging population that requires more and more attention.

The discussion about whether we want to ensure free, quality health care for all of us, including our elderly and poorer individuals, is political. No matter what we decide, health care will undergo a profound transformation closely linked to IoT technologies. There is no need to talk about it in a future tense as there are already advances in the area of medical diagnosis and remote patient self-monitoring.

For illustration, we have had automatic injection insulin pumps for years, although their use is not yet widespread. The device in question consists of a glucometer connected wirelessly to the telephone and a pump that regulates the administration of rapid-acting or short-acting insulin 24 hours a day. The cartridges must be replaced from time to time, as well as the entire pump. It may not be the most remote device, as the pumps are attached to the patient—who operates it from his or her phone or directly on the panel. Soon doctors will be able to monitor and intervene remotely in the administration of insulin to the patient. And this is no small matter: the prevalence of diabetes has doubled since 1980, from 4.7% of the world's population to 8.5%. In addition to its mortality, diabetes will be one of the most significant cost vectors for health care during this century.

All this comes hand in hand with the concept of telehealth or telemedicine—clinical services, such as diagnosis and monitoring, in remote. This type of service utilizes devices ranging from portable meters and sensors to a wireless conversation and monitoring systems that allow remote consultation to primary care centers.

Sometimes we find the term telemedicine separated from the concept of telehealth, which includes non-clinical services such as preventive care. Some of the devices designed for exercise, such as FitBit, would fit this definition. However, these users are generally healthy rather than sick, so they do not directly attack public spending—although they do prevent it. The World Health Organization uses the term telemedicine to describe all aspects of health care, including preventive care.

The breadth of disciplines in telemedicine today, still in its infancy, is vast. In developing countries, the focus is on mobile health or mHealth, i.e., the use of the mobile phone for remote medical assistance. Its success stems from some constraints, such as high population growth, high prevalence of diseases, low infrastructure—particularly in rural areas—and health specialists. And, on the other hand, the opportunity for rapid penetration of mobile telephony. With the arrival of 5G—which we will explain in a few pages—we will see more frequent cases of truly remote telesurgery, with the inclusion of surgical robots. These robots already exist for many applications but are always supervised and accompanied by a doctor. In January 2019, the first remote operation using 5G was reported, a liver lobectomy (liver removal) performed on a lab animal. It had been two decades since the first fully remote operation had been performed on a human patient, a cholecystectomy (gallbladder removal) performed by French doctors from New York and with the help of a ZEUS robot. This surgery is known as the Lindbergh operation and was a minimally invasive procedure. Unfortunately, the latency problems of pre-5G technologies have prevented further development of remote surgery.

The giant Medtronic acquired Cardiocom in 2013 for $200M USD, the first major acquisition of a telemedicine company in the world. More and more clinics and hospitals are relying on systems that allow health care personnel to monitor on an outpatient and non-invasive basis. In the United States, Synapse,[6] produced by doctorondemand.com (Dr+), was awarded in 2019 as the best telemedicine platform. Dr+, founded in 2012, has also signed several alliances with device manufacturers. If you are concerned about your health, you can schedule a visit through the platform, and use any associated devices it has, such as a simple thermometer, which will transmit your data before the appointment. Many people avoid visiting the doctor because of scheduling issues, because they cannot afford it, or simply because they feel too sick. Telemedicine reduces these challenges. Dr+ offers flat rates for patient visits, which helps people budget if they are uninsured, an essential issue in countries with partially or fully privatized healthcare. In those countries, this can mean the difference between a patient's life and death. In countries with universal public health care, it can mean the safeguarding and future sustainability of the system.

The cyborg movement holds the radically futuristic version. The first organization dedicated to helping humans become cyborgs was founded in Spain in 2010 by Neil Harbisson and Moon Ribas. Moon has a seismic sensor implanted in his arm that allows him to perceive all the earthquakes in the world in real-time. Harbisson is completely color blind from birth—he is only able to see in grayscale—and to perceive colors, he uses an antenna integrated with his skull through the occipital bone. This antenna transforms the color into sound frequencies, which he hears. It uses an internet connection, which allows him to send tones to other people through the network. "I don't feel like I use technology, I don't feel like I carry technology, I feel like I am technology," he says.[7] Does that sound like science fiction? With no need for surgery, the market for smart pills has been growing for years. Perhaps its most spectacular application is medical imaging. Like those of the Israeli company Given Imaging, acquired by Covidien in 2013 for 860 million dollars, then sold to Medtronic for 42 billion in 2015. They devised the PillCam COLON 2 (in the figure). It takes pictures of the intestine and sends them remotely to a nearby device, usually attached to the waist, and then to a computer for medical review. With this device, many patients avoid a colonoscopy.

Figure 5: Medtronic's PillCam COLON 2. *Credit: Medtronic.*

SMART CITIES

"A void like Chernobyl. A ghost town."[8] Thus, Shim Jong-rae described the Korean city of Songdo, the first Smart City in history, started in 2003. His testimony shows the intricacy of this type of undertaking and planned cities. So far, the city has cost the South Korean government approximately 40 billion euros. Today about 70,000 people live in Songdo, while the plan estimated a population of 300,000 by 2020. Only 1,600 companies have offices in the city and 58 are foreign. The costs are very high, and everything is still centralized in Seoul, which can be reached in about two hours by public transport, despite being only about 50km away. The city is not attractive to most people, no matter how much technology it utilizes, and its prices are prohibitive. *Gangnam Style*, a song that criticizes the elitist lifestyle of the Gangnam district of Seoul, portrays the city in its video. Songdo's choice for the clip is not accidental. It was dubbed *the ghetto of the rich* by the French newspaper Le Monde, in an article published in 2017. For more than a decade, urban planners studied the construction of Songdo carefully. The vision was to create a place free of cars, with wide green spaces and tens of kilometers of bicycle lanes—a paradise. But today, the streets, paths, and alleys are half empty. There is no cultural presence, no museums, nor theatres. The human touch is missing; the technologists' planning invisible hand can be perceived. Technology is ubiquitous, but on weekends, Songdo's few inhabitants escape to enjoy themselves in Seoul.

The first page of Spanish architect Miguel Fisac's *The Urban Molecule*, written in 1969, reads:

> *Our cities are sick. They do not work. They were made for centuries to be lived and lived in a different way than today. The massive demographic growth, the absenteeism of the countryside, the industrial concentration, etc., have hypertrophied some cities, which are on the verge of collapse. Others are growing vertiginously, and many towns are becoming deserted... It is urgent that urbanistic theories be structured, not utopian*

and for a distant future, but possible today and that they can guide, with real and accessible bases, the projects of remodeling, expansion, and even creation of new cities for the near future.

This reasoning is particularly valid for smart cities. Songdo, aside from a failed attempt at something that will undoubtedly work in the future, teaches us that the human factor is fundamental when it comes to technology. It is a looking-forward museum. It reflects, frustratingly, what cities can become: apartments with the latest trends in home automation, computers integrated into the streets to control the flow of traffic, interconnected condominiums where neighbors hold conversations by video conference. Everything can be done remotely, from opening the front door to attending university classes. No garbage trucks circulate; the garbage is vacuumed from the houses and recycled to generate electricity.

In the future, cities will develop in a similar way to what is being piloted in Songdo, more cost-effective, and with a human touch. They will be built on a hard foundation of automation and M2M technologies: efficient public lighting which is already happening in several cities; promotion of clean energy—Oslo has a plan to reduce its emissions by 50 percent by 2020 and 95 percent by 2030—whether renewable or high-efficiency cogeneration plants; further development of metering sensors, to enable, for example, automated parking payments or retail environments with smart shelf technology. All these cases also assume efficient and intelligent administration of the electricity grid: electrical substations to transfer from high to medium voltage, transformers, inverters, and power lines. Many cities, especially in seismic areas, have problems nowadays hiding electrical cables from the streets. It will be a particular challenge for these cities.

The city of Barcelona is a European reference in digital municipal management. It has been a pioneer, for example, in tackling droughts. Together with the companies Logitek and Wonderware, Barcelona has developed an intelligent system of city sensors for irrigation, which has led to a saving of 25% in water consumption in parks and gardens.[9] The sensors in the soil analyze the environmental humidity together with the expected rain level and modify the behavior of the sprinklers to help conserve water. The city has also made its Sentilo sensor and actuator platform available on the internet by making the source code accessible on Github. This allows urban planners around the world to study data from the smart city projects in Barcelona and Terrassa and learn from them.

One application will stand out from all the others in the short term: traffic management. Pittsburgh—which has in Carnegie Mellon University a pioneer in the deployment of autonomous vehicles—uses an intelligent traffic light system called Surtrac (rapidflowtech.com/surtrac). Surtrac is a system based on robotics and artificial intelligence techniques. It has cameras installed in public lighting and a local control system that sends optimized signals to its neighbors. In other words: the traffic lights *talk* to each other, ratting information about the traffic to reach the most optimal consensus possible. Surtrac processes information every second and sends information to connected vehicles and other traffic actors.

CONNECTED AND AUTONOMOUS VEHICLES

There is a misconception about where we lie with self-driving or autonomous cars. The Society of Automotive Engineers (SAE International) defines, in a multi-level classification, what an autonomous car is. We start from a zero level without any autonomy. The first stage of automation implements some driver assistance, where the vehicle takes control at times, as in the ABS braking system.

The second is partial automation of driving, where the vehicle takes control of the steering and brakes but is always assisted by a human being. Currently, we can find on the market many models with SAE 2 level, such as the Mercedes-Benz E-Class, the Volvo XC60, the SEAT Ibiza and León, the Volkswagen Golf, and the Audi A3, among others. Supposedly, level 2 vehicles require the driver to be attentive at all times. Despite this, Tesla level 2 commercial cars had already caused five deaths by the beginning of 2020.[10]

The third level is the one that allows the driver to fall asleep or move away from the control position temporarily. Despite this being an experimental level, we have one fatality: Elaine Herzberg, who died on March 18, 2018, in tests carried out by Uber while riding her bicycle. An example of level 3 can be found in the Audi Traffic Jam Pilot that implements the Audi A8, not commercially available, in any case.

Level four enables the driver to completely abstain from the driving action under certain measured and calculated environmental circumstances. The car can be operated without human intervention or supervision, but only under certain conditions defined by factors such as road type or geographical area.

The final level would give full control to the vehicle under any circumstances. Even a steering wheel would not be necessary. The only task the driver would have to perform would be to inform the car of the destination.

The reality is that the cars on the market today have not reached level 3. In the early years of the 2020s, we will see the first cars with level 3 technology on the streets in restricted areas. However, we are far from having completely autonomous cars driving on any street in any circumstance.

The main reason is that our infrastructure is not prepared for an autonomous car, so the "measured and calculated environmental conditions," which would allow the driver to escape, are never met. In May 2018, MIT presented an autonomous vehicle that can be driven without the need for maps.[11] This is a particularly impressive development because we rely heavily on the road infrastructure for these vehicles to function accurately, with highly visible lane markings or well-detailed maps for safe navigation. If we do not have this infrastructure, we depend on technologies that are not yet ready for most of the world's roads. Relatively simple impairments, such as potholes or lanes not suitably marked with paint, ruin the possibility of an autopilot because the car is *reading* these markings.

The design of cities that take this into account is essential. Waymo—a Google company—is already testing level 4 cars without a safety driver on the streets of Phoenix, but not all of them comply with the infrastructure that this city has. During the 2018 Las Vegas CES (Consumer Technology Association) meeting, the company Lyft made available to participants rides in a self-driven taxi, where the car circulated only on the streets, with an engineer as co-pilot and a security driver who only intervened in specific situations.

Precisely, the amount of distance a car can drive without human interruption is a metric used to measure the advance of technology. Until 2018, the Google car had managed to drive 18 kilometers in a row. It was followed by GM's with 8. Most of them did not make it to the kilometer mark. That same year, a 500-horsepower racing car managed to complete a 1.8-kilometer lap without any assistance in the United Kingdom. The video can be seen on EuroNews.[12]

There is something beyond technology when reviewing the abilities of self-driving cars: governments will need to take several important decisions in our transition to autonomous vehicles. For instance, under what atmospheric conditions should governments allow vehicles to operate at the highest levels of autonomy? They will also have to design how autonomous vehicles will live and coexist safely with traditional cars. One possible solution would be specially designated lanes. In any case, it seems that we are decades away from that.

In contrast, the concept of the connected car refers to how vehicles and infrastructure are connected by M2M systems to obtain benefits, such as increased safety—purely IoT, and more accessible. The autonomous car requires large doses of artificial intelligence; the connected car is simply another device connected to the internet.

The concept was first introduced to the market in 1996 by General Motors' subsidiary, OnStar. Today it is possible to purchase an M2M device for our car, connect it to the ODB port, and enjoy some remote applications, such as knowing our gas consumption or whether we drive aggressively or smoothly. Even without installing a specific IoT device, the real-time traffic information we get from applications like Google Maps or Waze follows the philosophy of the connected car. Furthermore, several brands have their own systems, such as Audi, Toyota, or BMW. These applications collect information and issue route recommendations that change in real-time according to the traffic we will encounter in a given period. Seem modest compared to autonomous cars? Connected cars are going to change our lives in many ways—for example, how much and how we pay for our insurance. Several insurance companies are already experimenting with charging models linked to our driving style, which they monitor thanks to devices with built-in accelerometers. Companies like the startup Vinli are already evaluating these metrics, and rating us on our recklessness behind the wheel. Soon, putting the car at high speed will cost us dearly, even if there are no radars around.

IOT-ENABLING TECHNOLOGIES

Several problems need to be solved to get the Internet of things off the ground. Let us take a quick look at some of them and what technologies will tackle them.

The first problem we come across is numerical: the connected devices must have some identifier. In today's Internet protocol, these are known as IP addresses. With the format we had been using (IPv4), we ran out of all available addresses a few years ago.

IPv4 had an ambition problem when it was designed. The protocol provided for the assignment of a host address to each machine that connects to the Internet, in the same fashion that when we send a postal letter, we include the full address of the recipient. These are the ones we connect to when we access a computer on the Internet. For example, when we type google.com into our browser, we are accessing the IP address 216.58.197.78. We can open a browser and type in those numbers, and we will access the Google page. A system called DNS (domain name server) is in charge of translating between the names we know to their pertinent IP addresses. So, when we are interacting with other computers connected to the Internet, we are constantly sending and receiving information from machines with assigned IP addresses.

An IPv4 address is composed of four blocks separated by dots with numbers ranging from 0 to 255, or eight bits. For example, the one we just saw: 216.58.197.78. One bit, remember, can take two values, 0 or 1, so it is capable of representing up to two different numbers. Two bits combine to take four values: 00, 01, 10, 11, so it can represent four numbers—for example, 0, 1, 2, and 3, so that 10 in the binary system is 2 in the decimal system. Similarly, to be able to represent the number 256, we require eight bits (2^8). Inasmuch as we have four different positions with the same possibilities, we will be able to obtain 2^{32} different addresses

($2^8*2^8*2^8*2^8$). This means 4,294,967,296 addresses. It seems like many possibilities, but the reality is that they were exhausted by the year 2011.[13]

Not all these addresses are occupied by unique machines. Specific ranges are reserved for special types of addresses, such as local IP addresses, for instance, 192.168.0.1. Hierarchical subnetting and virtualization have allowed the Internet not to collapse despite the lack of new addresses. But the implementation of an extended nomenclature becomes essential.

A new version of the Internet protocol, IPv6, will very soon replace the one we have. In fact, it has already been in use for some years. (If you are wondering why it is not called IPv5: a series of experimental protocols defined in 1979 and known as the Internet Stream Protocol, takes that name). IPv6 addresses are composed of eight four-element hexadecimal blocks, equaling 32 positions. Each element takes a value from 0 to 9 or from A to F, which totals sixteen different values. A single block of an IPv6 address has as many possibilities as a full IPv4 address. Having eight blocks, we have 16^{32} or 2^{128} possibilities, 340 sextillion addresses. It seems that they will be enough this time. According to a report presented by Gartner in 2018, 25 billion devices would be connected to the Internet by the year 2020. This is a staggering estimate, considering that the same report states that 4.9 billion devices were connected in 2015, which equates to a 400% growth in just five years. This increase sheds some light on how much IoT growth we can expect to see in the next ten or twenty years.

Once we have the devices identified, let us see how they can receive data. Commonly, when at home, we connect to a Wi-Fi signal issued by our router, which in turn is connected by a wire, be it copper or fiber. In domestic applications, devices utilize this local connectivity to communicate, or other short-range protocols such as Bluetooth. To have remote and distant devices, however, we cannot merely deploy fiber everywhere and install thousands of Wi-Fi routers as if they were breadcrumbs. Most often, devices have a SIM card installed, like those in our phones, and connect using the mobile phone network.

Do you remember remote surgical operations? The first pilots suffered from high latency, i.e., a low reaction capacity, which is unacceptable when intervening surgically. The acronym 5G refers to the fifth generation of mobile phone technology. Each new G marks a radical change in the nature of the data being transmitted over wireless communications networks, be it speed, bandwidth, or latency.

The first-generation network was analog, i.e., it transmitted the entire signal (not digitized) and only allowed voice calls. Not all networks were based on the same protocol, making it a rather heterogeneous generation. Despite that, this technology was still in use until 1990, and the world's telephone stock was around 20 million. The second generation (2G) was already digital technology and under a standard protocol: GSM. With 2G it was possible to send text messages for the first time.

The introduction of 3G drastically increased the data transmission rate and its capacity. In addition, it provided multimedia support, allowing video and photos to be sent by message—the failed MMS—and integration with the Internet, TCP, and IP protocols. The fourth generation increases the bandwidth and reduces the cost of resources.

The importance of 5G will not be small. Home automation and connected appliances help to automate household tasks, which not only improves personal comfort but also helps those who need assistance with everyday tasks. Nevertheless, the significant advances in autonomous vehicle technology or remote surgery that we have seen before are only possible with 5G. For many people, the increase in speed will be the most visible feature of 5G. However, it is imperative to note that the technology promises to provide other benefits, from improved responsiveness, by reducing latency, to the ability to connect more devices at the same time. Lower latency means greater reflexes of connected objects, which is key to many of the real-time applications and will become critical as more and more people make increasing use of smart devices. 5G promises to enable IoT for use with 3D holograms, virtual and augmented reality, intelligent cities connected in public spaces, better traffic control, and many other applications that rely on near-instantaneous response time.

With more intense use will come more battery power problems. We already suffer from them with smartphones, whose battery capacity has not scaled up in the same way as their power, so they no longer last as long as the first models. To a large extent, the IoT's ability to work in hard-to-reach locations will depend on how the batteries evolve over the next few years.

Fortunately, this problem is getting the attention it deserves, with researchers around the world working hard to address it. New materials are being devised, experimenting with aluminum batteries to replace the lithium-ion batteries used today, but with the possibility of collecting energy from Wi-Fi and Bluetooth radio waves.

The best solution in the short term could be *to have no battery*, with the evolution of communication standards towards low consumption such as ANT+, Zigbee, Z-wave, and even some longer-range ones like Sigfox or LoRa. Over the last decade, the deployment of battery-free applications with power harvesting had only been viable for very short-range applications such as NFC. But let's take the example of Bluetooth 5. It has the lowest power consumption of any of the previously implemented wireless connectivity solutions and allows you to collect power from your signal. Bluetooth 5's range has increased four times and is now comparable to Wi-Fi. Therefore, the battery life of an IoT device connected to Bluetooth 5 has been significantly improved. With the new standard, a radio reduces its power consumption by five to ten times. Considering the projected number of IoT devices in the world, this reduction has a huge impact on global energy consumption. With ultra-low power technologies, power consumption is low enough to be supported by radiofrequency energy, extracted from the signal itself, but also light or heat collected by photovoltaic or solar thermal panels. In other words, for many IoT devices, the power consumption could be lower than the energy that can be collected by the device.

We haven't come to this yet. At the moment, we know that our battery consumption trajectory is not sustainable. Frequent battery changes are not feasible. We must and will find ways to increase the battery life of connected devices, similar to the transition we saw in incandescent bulbs to LED lights. Consumers themselves will demand this change.

In the same way that batteries will evolve—eventually disappearing from some applications—the SIM card we use in our phone has undergone some modifications since its introduction in 1991. From the size of a credit card, the SIM evolved to the classic miniSIM (2FF) and started to shrink, first to the microSIM (3FF), then to the nanoSIM (4FF). The miniSIM arrived in 1996, but it was not until the launch of the iPhone 4 that we moved to the microSIM, and the iPhone 5 when we leaped to the nanoSIM. And we will soon forget about all of them because the eSIM or virtual SIM has arrived to replace them.

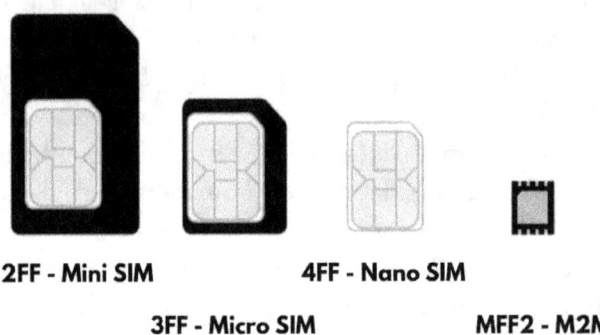

2FF - Mini SIM

3FF - Micro SIM

4FF - Nano SIM

MFF2 - M2M

Figure 6: evolution of the SIM, from standard to e-SIM. *Credit: Hologram.*

The eSIM concept is based on an integrated SIM called MFF2, sealed directly on the circuit board during manufacturing. From a technical perspective, it works in the same way as a normal SIM card. The fact that it is integrated, however, makes the MFF2 a perfect choice for devices in difficult conditions, such as being utilized in outdoor conditions (it is protected from the weather) or in constant movement and vibration. The chip is also permanently implanted in the device and is not duplicated or removed in the event of theft, making it safer—except in a situation where the entire device is stolen.

It is important not to confuse the MFF2 chip with the eSIM solution. eSIM is a software protocol that is implanted by soldering an MFF2 onto the circuit board, but could also appear on a 2FF, 3FF, or 4FF chip. eSIM should be considered as software, not just a new type of SIM card. It follows a new standard, which brings many benefits to the table, along with some doubts. For example, how are we going to change the mobile operator if it is not possible to extract the eSIM? Changing service provider will be even simpler since it will be enough to tell the operator that we are going to change the numbering of the eSIM—an ICCID code of 19 or 20 digits—to associate it with the new provider. We will also have the possibility of associating it with more than one operator from different countries. The eSIM will make it easier to link the same number to different devices, and it could also make it easier to have a single invoice and a pricing plan for all of them. The solution is deployed once and modified remotely for decades without compromising security.

We may no longer need a phone to communicate in the future: for many users, especially younger ones, communication is based on messages from WhatsApp, Telegram, and other applications. On the other hand, until now smartwatches received data from the phone via Bluetooth, but with eSIM, they can have their separate contract and connection. Many users would be more interested in purchasing a smartwatch and wireless headset for less than a mobile phone and using it to call friends and listen to Spotify.

But even before entering the end-customer market, eSIM is already being used in both large devices with longer life cycles (like vehicles) and smaller, often portable devices with relatively short life cycles, such as the Apple iWatch. The consumer electronics industry is expected to lead this market in terms of the number of connections soon. Deutsche Telekom launched the nuSIM at the end of 2019, specifically designed for low-cost devices used in long-life IoT mobile applications, such as asset trackers or intelligent motion or temperature sensors. The automotive sector is expected to generate the most connectivity revenue. Utilities, transport, and security sectors are also moving towards adopting eSIM as the standard.

All indications are that from 2020, eSIM will start to be included in mass and default in the latest high-end smartphones. We will even see the expansion of eSIM in the portable and convertible sector. There could be a transition period where we see both an eSIM and a nanoSIM slot in smartphones, giving operators time to adapt to the virtual card, but the arrival of eSIM is already a fact.

BLOCKCHAIN

Strictly speaking, blockchain is not an IoT-enabling technology. I wanted to include it in this section because of its unquestionable interest and some synergies among both technologies that I will comment on at the closing. Let us briefly understand how it works.

First, there was bitcoin, which is a cryptocurrency. Here begins the confusion. Bitcoin or blockchain? The 2009 bitcoin-founding paper, written by Satoshi Nakamoto—who, despite the name, is believed to be a group of people of non-Japanese descent—never mentions the word blockchain. Bitcoin was a brilliant proposal to solve some of the problems of creating a non-centralized currency using a peer-to-peer network. It acts as a network similar to the ones we talked about a few pages ago, Napster and eMule. Shortly after the paper was published, it became manifest that their elegant solutions and data structuring were also useful in other applications that required an accounting record or ledger. The concept of blockchain was born, and it started to become popular from 2015 onwards.

The common use of the article *the* just before the word blockchain seems to imply that blockchain is a specific instance of something, a material thing, visible and palpable. Still, it is not. Blockchain is an idea of data structure, an abstract guide to store information, implementable with your favorite programming language, or by borrowing pre-existing code in any language. JavaScript is the most common language used at the moment. There are several possibilities and a certain degree of freedom when writing a blockchain.

The most basic implementations found on the net have less than 200 lines of code.

So, blockchain is a data structure. What does that mean? Merely, a data structure is a way of storing data, accompanied by some useful rules for executing operations on what is stored. For example, a shopping list is a straightforward form of data structure. It seems obvious, but structuring it as a list has some advantages. By separating each object into a line, we recognize that there are different objects. If we received a shopping list in Chinese, we would be able to identify those objects even if we did not speak a word of Chinese. Word by word, or logogram after logogram, we may try to translate it. If the objects all appeared in a row, it would be impossible for us to understand the message. We would not be able to distinguish if we were dealing with a complete sentence or a separate list of items. The structure gives us additional information. A spreadsheet table, like in Excel, is another type of data structure, each cell identified by its column and row. An alphabetical list is a data structure with some rules: there is an order between the items; in this case, the first letter of each word.

Blockchain, in its most unadorned form, is a succession of blocks, each containing an identifier, a timestamp, some data, and a reference to the next block.

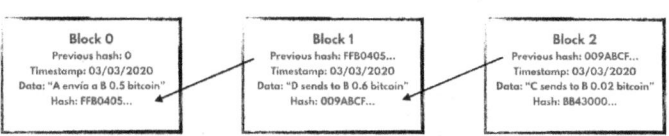

Figure 7: a very basic blockchain.

We could generate a very rudimentary chain of blocks on paper or our computer, by merely opening the word processor and producing text files with an identifier — 1, 2, 3... — the date and time when we created each new document, some text, and a reference to the next one. It is that simple: a bunch of text files (each one representing a block), including data, timestamped and linked to the previous one. We can use this structure for many things. In the case of bitcoin, the stored information is, of course, monetary transactions between pairs. Any block in the Bitcoin string will include messages like "person A transferred 0.3 bitcoins to person B."

That does not sound impressive. We need to add more stuff, and the first task is to distribute the information. We send a copy of the entire chain to all the nodes in a network. When we open a wallet to invest in cryptocurrencies, we first download the whole chain of transactions from the start of bitcoin to our computer. They are there, all of them, stored. Remember: one of bitcoin's objectives was to create a currency without the intervention of any centralized institution—a central bank, like the FED or the ECB. Let us think of blockchain as one variety of distributed and decentralized databases. The difference between the two is that in blockchain, each participant has a complete copy of the entire record. Having the whole accounting record at each node in the network makes the blockchain different from some other distributed database implementations, such as MongoDB or Cassandra, or any of the new database paradigms. The Apache Hadoop distributed file system (HDFS), which we will discuss a little later, is another type of distributed file storage system.

What makes blockchain different is that it is an append-only structure. You are not allowed to modify or delete any old records. It is only accepted to correct by adding new blocks to the chain. The fact that we cannot alter the past is vital when the participants do not know each other, and it is necessary to generate trust between peers. Do not forget that even if you are transferring money to an identified person, you are keeping the record with a community of people who are not.

Now, I imagine you have some doubts. For example, "If I download the entire bitcoin transaction log onto my computer, does that mean I can spy on what everyone has done with their money in the past?" The truth is that you can see these transactions, but that does not mean you can understand them. All bitcoin transactions are encrypted, ensuring privacy between users. The way bitcoin encrypts its transactions is through the application of hash functions. A hash is a mathematical function that converts any randomly-sized information to a fixed size, usually an alphanumeric string—something like "e0bc42f." More specifically, Bitcoin uses SHA256, which converts data into an exactly 256-bit string, represented in 64 positions containing a number between 0 and 9 or a letter between A and F. For instance, let us try to convert a couple of sentences to a SHA256 hash:

What we talk about when we talk about innovation
8CCDF3A8195797582D4EC44D308606BB3BC07EC6C124CDAAF528A2D4151EF6C1

Hello!
334D016F755CD6DC58C53A86E183882F8EC14F52FB05345887C8A5EDD42C87B7

Think of a hash as a signature or a security seal. It is a smart way of ensuring that no one can manipulate an old block in the blockchain. The hash is the result of putting all the data of a block into a blender that returns a unique alphanumeric string. The way blocks are related to each other is not through simple numbers, but utilizing hashes. The hash is the result of applying a formula to the data contained in the block. If I alter anything, the resulting hash will be different. If I try to manipulate a block by changing the time it was created, the resulting hash will change. If I decide to change the amount of money transferred, the hash key will break. The same is true if I try to manipulate the order by saying that the previous block is different. The hash chain guarantees that the story is incorruptible.

But let us suppose that I want to falsify a five-block blockchain by altering some data in the second block. The resulting hash will be different. Nothing prevents me from obtaining the new hash and then editing the third, fourth, and fifth blocks until I have a well-formed chain. With enough patience, I could fake the whole chain and achieve my goal. Here is where the proof-of-work comes in.

The proof-of-work is a mathematical puzzle that requires time and computing effort for any given node to add a block to the chain. In the case of bitcoin, the problem consists in adding a number, called the nonce, with the condition that the resulting hash starts with a predefined number of zeros. In other words, we tell our CPU: "Here is this block data, now combine it with something until the resulting hash starts with, say, thirteen zeros." It needs to start testing and testing. In the case of bitcoin, the difficulty is adjusted so that it is completed in an average of 10 minutes, and it changes over time as computational capacity increases. People who dedicate computer power to these puzzles and help form the blockchain are called miners. They are rewarded with new bitcoins each time they add a block to the chain. This is the only way to create new bitcoins, just as central banks print currency. The difference is that the bitcoin creation rhythm is well-known, and will be until reaching 21 million, from which no more will be produced. That is why bitcoin is said to be a deflationary currency. Soon, new coins will no longer be minted.

Let us now return to our attempt to manipulate the five blocks. It will be difficult for me to achieve my forgery if I need to spend an average of 10 minutes on adding a block in a distributed network. Anyone is allowed *a priori* to add a block. Once I manipulate the second block, I would need to be the fastest at adding the new third block, and again at adding the fourth block, and again at adding the fifth block. If the network is big enough, that can't happen. This is how you prevent fraud in the blockchain.

Blockchain, therefore, allows a group of unknown people to keep a ledger together, without the need to trust each other, only with the warranty of the security mechanisms that the algorithm itself implements. There are also private blockchains, used in the internal scope of organizations, where all this is not necessary, and the security is configured through a process called *selective endorsement*, where authenticated users verify the transactions. In blockchains, where the participants do not know each other, the work test ensures a system of consensus and mutual trust.

Why will the interaction between blockchain and IoT be so relevant in the future? The centralized scheme on which IoT is being driven will not be viable in the approaching ocean of connected devices. The servers' capacity will be overwhelmed and become a source of security risks. We will require to increase the investment in cybersecurity, making business cases for some IoT applications impossible. We will be thus challenged to provide enhanced autonomy to the devices: energy independence, employing self-sustaining applications that do not require batteries, and decision-making sovereignty without central supervision. The trust generated by structures such as blockchain adjusts well to a scenario of millions of connected devices interacting and guarding the interests of different people or companies. By using smart contracts, many processes will be automated, and the devices will have the freedom to make decisions without going through a central server.

All of this, in reality, is not so futuristic. Several companies have been implementing it for some years now. The Danish logistics and transport giant Maersk has been implementing

IBM's Hyperledger solution to change its global supply chain for some time.[14] The Australian telecom Telstra is investing in blockchain projects with IoT to understand how to protect its intelligent home devices from attacks. They are combining biometrics, blockchain, and IoT to detect any security blackhole, by attempting to hack themselves the networks connected to the IoT devices.[15] The Finnish innovation hub Kouvola also uses IBM technology to combine the two in a project funded by the European Commission.[16]

BIG DATA

Although the expression *big data* is becoming popular during this century, it is at heart something antique. At all times, we stored data. At all times, experts felt they were dealing with huge amounts of it.

The disciplines that today relate to big data are nothing more than a sophisticated evolution of Statistics. Since 1663, the first rudiments of analysis have appeared: John Graunt in this year recorded and analyzed data on the bubonic plague in Europe to warn about its consequences. What we call "big data" today differs in the tools, but not in the objectives. It involves the creation or capture of large amounts of complex data, its storage, retrieval, preparation, and finally, its analysis.

The first record of the term "business intelligence" was coined in 1865 to describe the competitive advantage that the banker Henry Furnese had obtained from analyzing relevant information about his financial activities. The term began to become popular in the 1950s. In the mid-1960s, the first data centers were born, such as IBM's in the United States, designed to store fingerprints and American tax information. In 1970, Edgar Codd described relational databases, which we will discuss in a moment. During all these decades, we find numerous references to the concern of employees in the new computer disciplines regarding analyses and processing of all new information. What differentiates the first rudimentary statistical and demographic analyses, the business intelligence of the 20th century, and the arrival of the modern "big data" is simply a matter of magnitude on several levels.

First of all, these concepts differ in volume: the amount of data generated is growing exponentially. Although we often cite the typical cases of social networks, there are examples everywhere. In 2009, the Indian government launched Project Aadhaar, which records biometric information—iris scans and fingerprints, among others—for residents of the country. By 2019, it had registered 1,246 million individuals, making it the world's largest biometric database to date. Each of Boeing's aircraft generates ten gigabytes per second on each flight. The Internet of Things produces colossal amounts of data every minute, and so on.

Although its quantity is the most vivid facet of data, so is speed and variety. Let us think about the operations carried out electronically, every second, in the stock exchanges, or the magnitude of the information generated by the new online versions of the most popular video games. The data processed includes text, audio, video, social networking information, and much more. It is both structured and unstructured.

What does "structured" and "unstructured" mean? Most professional databases in recent years have been designed under a mathematical paradigm called the entity-relationship (E-R) model, described by Edgar Codd. The E-R model is an abstract representation of reality, in which "entities"—objects or people—"relate" to each other through associations. Entities also have attributes that describe them. For example, an entity can be attributed with the individual's name. Multiple relationships can also contain attributes. In the relationship between an "order" and an "item," there may also be a "quantity" of them. Even if you have never seen a relational database, I am sure you can understand this entity-relationship diagram in figure 8.

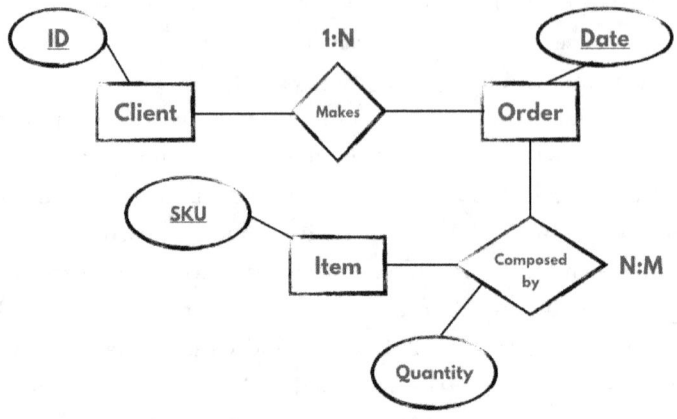

Figure 8: example of an E-R model.

The diagram describes a business relationship. A customer, who has an ID card as a defining attribute, can place a certain number of orders, which are identified by their date. In turn, an order consists of some items of a specific type, which we discover by their stock-keeping unit. Vice versa, an item can also belong to several orders, because they are not unique. It is a diagram that models a simple commercial activity. How could it be improved? A possible improvement involves expanding it to find out how many items we have in stock so that we can avoid ordering more items than we have. As it is designed, there is no "warehouse" entity. (We are often surprised by software that does not allow a particular action. Remember that its developers have to start somewhere, modeling step by step). This model we have created acts as a mold that accurately describes what information we should record and how we should record it: a client with his ID, an article with its serial number, and so on. This is called structured information. We define first the data structure. Then, we populate our database accordingly. Relational databases with structured information have dominated data management until the beginning of the 21st century.

With the rise of colossal amounts of unstructured data, it has become necessary to build non-traditional computer applications for the appropriate processing and treatment of these vast and disordered bodies. We call the architects of these applications data scientists.

Now let us see what it means to deal with unstructured information. For anyone unfamiliar with a relational database, an intuitive way to think about structured data are personal to-do lists. Let us include the date and the person to whom I have to deliver something. You could think of an E-R model: a task will have as attributes an identifier, a description, and a deadline. A person may have a name and a position. Imagine it even simpler: an Excel file, where each cell occupies a specific piece of data of one particular type. Let us have all the delivery dates in one column. I can run actions on them, like sorting, so I can remember what the most urgent task I have is. Remember the example of the Chinese list we referenced a moment ago? The structure provides information by itself, and software using that data structure may provide extra functionality.

I could hold the same information in plain text format, or on a piece of paper, handwritten. The information may be the same, but it is no longer structured for a computer—just text. I have no way to make my computer understand that 12/04/20 is a date because it only understands that there are six numbers and two sidebars. It cannot understand my handwriting when I scan the handwritten paper. However, I could show it some tricks: "When you see an expression in the form dd/mm/yy, that is a date. The first two numbers represent a day, etc." That can be coded. Programmers use such techniques in text searches when implementing regular expressions or for demarcating network segments when using masks. It is also relatively simple. But what if I wanted to show my computer what a name is? You could check it against a list of names I give to the machine and lookup for those words: John, Joseph, Rose. Propose some rules. Names are always written in capital letters. What if I find a file with a name in lower case? What if someone always writes their tasks starting the sentence with a capital letter? What if I come across a Tuesday or an August? Are these people's names? They are not.

This is just a glimpse of the challenges that data scientists encounter every day. Like paleontologists, who, faced with a dinosaur dig, stick pikes in the ground, mark out areas and carefully brush each bone until it is labeled, data scientists have to comb out, with fine brushes, data sites that are not labeled in any way.

Notwithstanding, some of the raw material that data scientists work with is not unstructured. Maybe 80% to 90% of it is. Even this information has some meaning and internal structure, and we sometimes relate to it as semi-structured data. E-mails have a sender, a receiver, a subject, and a body. Plain text documents have a language, and they cannot have a video embedded in them. Videos have a compression algorithm, like MPEG, which allows us to know certain things about their structure. Photos, audio files, presentations, web pages, and many other types of business documents as well are unstructured as a whole. The data they contain is not encapsulated under an E-R model, where there is a "Client" entity with a "last name" field that we can consult and from which we can expect certain types of specific information. Perhaps it would be more accurate to say that data scientists work with heterogeneously structured information. Their work is more like tidying up our room than teaching a computer what a name is.

In the last few years, specialized database paradigms have appeared to treat unstructured data, known as NoSQL. SQL is the primary query language in relational databases. These databases do not, therefore, follow the entity-relationship conceptual model.

There are several types of NoSQL databases: key-value databases, which link a data to a key; network databases, frequently used in social networks; and document-oriented databases, which differ from the former in that they structure the data by linking the key with a format: XML, JSON or BSON.

Big Data can also take advantage of the flexibility of the Cloud through the use of object storage in the cloud, with the use of the serverless paradigm, which consists of deploying code without worrying about the underlying infrastructure. We simply forget that there is a server and send our logic to a server that we ask it to run. In both cases, you pay for what you consume.

FROM DATA WAREHOUSES TO HADOOP

Now that we have a better understanding of some concepts, let us dig a little deeper into the history of the data. In the 1990s, most organizations had repositories called "data warehouses." These dealt with structured, historical, and normally organized information by subject in subsets called data marts. There could be, for example, the finance data mart. An element that defines a data warehouse, and ones that differentiate it from a normal relational database, are OLAP cubes, also invented by Edgar Codd. They arrange the data in vectors of various dimensions in order to facilitate the analysis of extensive amounts of data. This was necessary because in the 90s, the concept of Big Data didn't exist, but there were already considerable amounts of data to be analyzed.

We are not alone

Hadoop's arrival in 2006 was revolutionary for the development of Big Data and our ability to deal with unstructured data.

In the early 2000s, Google already had its search engine, its PageRank sorting algorithm, and its spider placed online that surfed the web and retrieved information on published web pages (GoogleBot) running. GoogleBot worked in indexing all the information that fell into its clutches. At one point, Google found that their work was piling up. There was not enough capacity. They were not indexing at the rate they wanted. It was also understandable to think no code — especially one implemented in the hardware of that time — could index the whole internet on a single machine. So, Google decides to expand the number of machines to four, which increases the advance. But they still need to coordinate the work of each machine manually, merging all the information.

At the same time, more people were concerned about the indexing problem, such as Doug Cutting and Mike Caffarella, who had just started working for the Apache Foundation. One fine day in 2003, they discovered that Google published an article[17] in which the authors described and solved some of their problems related to data storage. They implemented it in Java, giving rise to the distributed file system NFDS — from Nutch, later renamed to HFDS, from Hadoop. Hadoop didn't exist yet. Well, Cutting and Caffarella have solved the structural problem. They will no longer use a single machine and have a layer of distributed storage with extra features: it automatically rebalances within the cluster, so that no computer is overloaded while another is idle; it can handle hardware failures; it does not lose data packets; and it is not rigid, like the relational database models we have seen before. Wonderful. But they still need a way to manage that data once it is stored. And what do they find? Another Google programmer that makes data scientist's jobs easier. In

December 2004, Jeffrey Dean and Sanjay Ghemawat published their article[18] on MapReduce.

MapReduce solves three things: how to compute information in parallel; how to distribute the processed data; and how to manage software bugs. And MapReduce does it elegantly with two functions: *map* and *reduce*. It frees the programmer from the tasks of distributed programming. In other words, MapReduce allows software that has been written in a common programming language to be executed in a Hadoop cluster without radically changing its code.

In 2006, Google publishes *Bigtable: A Distributed Storage System for Structured Data*.[19] That same year, Cutting joins Yahoo! along with the Nutch project. Finally, in January 2008, Yahoo! launched Hadoop as an open-source project for the Apache Foundation.

Distributed computing was not an unknown concept. In a certain sense, the packaging the TCP/IP protocol uses to transmit information over the Internet is a form of distribution in a computer network. The P2P networks we were talking about a few pages ago are distributed computing. The Hadoop story in comparison first translated those concepts into managing massive amounts of unstructured data.

Hadoop is therefore based on many small computers, each of which is responsible for storing, processing, and analyzing a portion of enormous volumes of data. The greatness of the system is that, although each of them works independently and autonomously, they all act together, orchestrated, as if they were a single computer of incredible dimensions.

BIG DATA APPLICATIONS

Making sense of the data, extracting and preparing it, are just the first steps. Big Data opens the doors to an amazing collection of practical applications: predictive models, clusters, artificial intelligence. Once we have clean data, the next step is to mine it. The goal of data mining is to extract patterns and knowledge from large-scale data sets, which can be reconfigured into a more understandable structure for further analysis. Probabilistic programming allows us to create learning systems that decide by inferring from these data and previous knowledge and apply it in the analysis of medical images, financial predictions, or atmospheric forecasts.

Statistics is nothing more than treating data to draw conclusions from it. All AI starts from statistics: from the descriptive one which visualizes the basic characteristics of the data being studied; and the inferential, which is used to draw conclusions beyond the visible.

The applications and uses of Big Data are infinite, but they can be categorized into two broad approaches: outwards and inwards.

Outwards means harnessing the tide of data to better understand your audience and act accordingly. You would segment your clients, design communication campaigns, and create products. We need not limit ourselves to thinking of a company. We find some of the most famous use cases in electoral campaigns. In 2016, Donald Trump's team paid a fortune to the late Cambridge Analytica consulting firm to analyze data from millions of potential voters. The same consultancy ended up mired in a scandal in 2018 when a non-consensual use of data from Facebook users was discovered. Cambridge Analytica admitted running campaigns to change the vote. Obama already utilized a similar analytics strategy - but with consensual use of data - in his 2012 election victory, with a slight difference. The Obama campaign focused on identifying undecided voters, i.e., segmenting its population to attack those who could change their vote in their favor. The Trump campaign decided which topics would work best to attract voters, for example, the controversial immigration on the Mexican border. With his particular style, Trump left neither side unmoved. This radical behavior allowed for a clearer and easier monitoring of citizens' reactions to their statements. Once the analysis was complete, the system sent personalized messages to 100,000 potential voters every day.

Inwards implies improving the way of working internally, based on similar analyses. As we move through a chartered territory, with a greater quantity of structured data, it is here where business intelligence has expanded the most, through preparing dashboards and statistics of waiting times, sales and others. In complex enough organizations, like those with thousands of points of sale or logistical interconnection, or a telecommunications company with thousands of radio bases deployed, developing analytics for these sizable amounts of data becomes challenging.

For both approaches, we can point to two broad data-mining families: classification and prediction.

A classification model will take a set of data and try to group it, labeling it according to common traits. We call this clustering. Predictive models will try to guess future behaviors. For example, predictive models will anticipate defaults for a group of clients before they occur. The difference between the two is not always clear, as nothing prevents us from classifying our clients who are likely not to pay as "risk clients." If the variables we have chosen are suitable predictors of future behavior (a risk client will end up not paying), this classification can serve as a predictor.

There is confusion regarding predictive models and their ability to anticipate events. These models appear to "know what will happen" and many institutions blindly indulge in mathematical models for making important decisions. It is likely due to these two factors: ignorance of the statistical background, which leads us to believe that these models are more accurate than they are; and the somewhat controversial use of the term *prediction*.

A predictive model, in terms of statistical analysis, tries to infer new variables or unknown characteristics of a phenomenon from traces or variables that are available. Most times the future is not *predicted* and in all of them, there is a certain margin of error.

Let us take the example of three systems that try to predict a person's height.

The first system is a method that pediatricians have long used to estimate the height of a young child from that of both parents. It is a simple formula, with a good enough margin of error, which finds the mean and adds or subtracts a certain amount depending on whether it is a boy or a girl. A lavishly populated mathematical model could improve that formula and make it more precise. This is a system that predicts the future.

The second system supports anthropologists and historians trying to find out the size of hominids from skeletal remains discovered in excavations. We have femurs, tibiae, fibula, humeri, ulnae, and rays, and we can know the sex of the individuals to whom the remains belong. With these data, we are looking back into the past, wanting to find out data that is unknown (the size) from other known data (bone measurements). This has also been going on for decades, with simple models using only the femur[20] up to models that break down and apply distinct formulas according to the bones used.[21]

A third system could try to combine the previous two to find patterns that relate, for example, fetal femoral length to the height of the neonate in adulthood—a relationship that at the moment seems unknown.

Many of the models that exist are of the second type. This is not a "prediction" in the temporal sense, as the weather forecast would be, but a statistical inference with a probability of being true with a margin of error—and not without risks. The same occurs with another type of well-known statistical study: electoral surveys. Since it is impossible to interview the entire voting population, a study is carried out on a smaller universe. These data are "cooked" at the time of known phenomena, for example, that a particular population is less likely to admit a conservative vote than a progressive one. Or a population that directly lies about the vote. We know that the weather forecast or electoral polls are not reliable. Why do we think analytical models are?

In most cases, when we read a paper or statistical study, the authors describe their methods somewhere well visible. We are informed of the margin of error to which they subject their statistical calculations. The confidence interval usually stands around 5%, with a confidence level of 95%. This means that if polls show that a candidate will get 46% of the votes, what it truly tells us is that they will get something between 41% and 51%, and furthermore, we are sure that this is true at 95%. They can achieve these levels with small sample sizes that agencies can deal with to produce polls. We experience something similar in newspaper headlines of the form: "study of the prestigious institution X shows that Y." The study in question should somewhere have its statistical model described. One of the most important data is the p-value, ordinarily limited to below 0.05 or even 0.001. We call that limit the significance value or alpha. In Statistics, the p-value shows the probability that an experiment will return a result "by chance." In other words, if a study limits its p-value below 0.05, it means that these results have less than a 5% chance of having occurred by chance. You may think that with such low probabilities, a study can boldly confirm what it is asserting. However, at a 5% rate, I could make up that I can hypnotize coins, so they would obey when tossed into the air. If I flip five coins in a row with a predicted result—say, five tails in a row—I will comply with my margin of error. The probability of five tails in a row is 0.03, while my p-value is 0.05. I could indeed affirm that what I am doing is "hypnotizing" the coin.

This is impossible. And yet, three times out of a hundred, it will come true. Is the occurrence of three out of a hundred so little, regarding the stupidity that I am affirming? Data is easy to manipulate. Statistical experiments are not infallible and, consequently, neither are our mathematical models. Therefore, a p-value of 0.05 is accepted in sociological research, while in medical research, where making an error can have serious consequences, one of 0.01 is used.[4]

Let us return to the case of the American elections. A well-known study from 2013, led by Michal Kosinski, has greatly influenced the work of consultants such as Cambridge Analytica, stating that it is possible to discover certain behavioral aspects from the behavior of individuals on Facebook, in particular, their "likes."

The study,[22] carried out on 58,000 volunteers, states that his model:

> ...*correctly discriminates between homosexual and heterosexual men in 88% of cases, African Americans and Caucasian Americans in 95% of cases, and between Democrat and Republican in 85% of cases.*

The analytical model attempts to relate behavior on Facebook with a widely accepted psychological theory called the Big Five, similar to another very popular model, the Myers-Briggs indicator, used in many organizations and schools. The Big Five model divides our psychological traits into five blocks, while the Myers-Briggs model breaks it down into four. Both use a questionnaire that the analyzed subject must answer. Kosinski asserts that we can obtain similar results with no questionnaire, through the likes that the subjects give to different web pages.

Some aspects of the study seem sharp, such as the fact that

---

[4] This is just a very brief summary of the weaknesses of statistical hypothesis testing.

> ...few users were associated with Likes explicitly revealing their attributes. For example, less than 5% of users labeled as gay were connected with explicitly gay groups, such as No H8 Campaign, 'Being Gay,' 'Gay Marriage,' 'I love Being Gay,' 'We Didn't Choose To Be Gay We Were Chosen.' Consequently, predictions rely on less informative but more popular Likes, such as 'Britney Spears' or 'Desperate Housewives' (both moderately indicative of being gay).

Others are puzzling:

> ...best predictors of high intelligence include 'Thunderstorms,' 'The Colbert Report,' 'Science,' and 'Curly Fries,' whereas low intelligence was indicated by 'Sephora,' 'I Love Being A Mom,' 'Harley Davidson,' and 'Lady Antebellum.'

When we make these types of models, we fall into the risk of linking variables that do not have a real cause-effect relationship. The divorce rate in Maine correlates perfectly with margarine consumption. The amount of consumed mozzarella cheese fits well with the number of doctorates in Civil Engineering in the United States. The US public spending on science and technology is related to suicide by hanging. And the number of people drowning when falling off a boat, with the marriage rate in Kentucky.[23]

"Correlation does not imply causation." The empiricist David Hume worked hard on this, arguing that we can never perceive the causal relationship, only the correlation. For a cause to exist, there must at least be a subordinate temporal relationship between an event and its cause. For example, mills moving cannot cause the wind if it has appeared before. The opposite case can occur, but it can also happen that they are mechanical mills and the wind is not the reason that they move. It is even a well-documented logical fallacy: *cum hoc ergo propter hoc*. To define the cause, we must run experiments.

We must treat with care all the conclusions drawn from data. Data is a human invention. We define the phenomenon we want to measure, design systems to collect data on it, clean and process it before analysis, and choose how to interpret the results. Even with the same data set, two people can reach different conclusions. It happens too often. This is because the data alone is not a fundamental truth. The data are observable, verifiable quantities that reflect reality. But the judgments derive from other realities, they are subjective — sometimes they have hidden interests. What people call data can be cherry-picked, prepared only to support an agenda; or random collections of information unrelated to the real world; or information that seems reasonable but comes from unconsciously biased collection efforts.

Sometimes we forget that citizens and clients are not data or numbers, but rather people with names, surnames, and a life. Sounds romantic these days. Bad analysis of a lot of data can lead to errors in interpretation. Few, very significant data about your customers can give you more clues than an automated analysis of thousands of anonymized data. Design thinking techniques are also based on this axiom, obtaining little high-quality information from a few potential clients, which you can manage in one session.

A ROBOT TOOK MY JOB AWAY

3

> *"I am so clever that sometimes I don't understand a single word of what I am saying."*
> —Oscar Wilde, 1854–1900

In 2006 Nintendo launched the Wii, our first contact with robots equipped with artificial vision and capable of interpreting actions and commands by gestures. In the industry, however, robots are old acquaintances. Industrial robotics is probably the most widespread branch of automation. It includes the use of machines to carry out manufacturing tasks, diminishing human intervention. Robots are just a subset of the long lineup of technologies, including motors, hydraulic or pneumatic actuators, sensors, machine vision cameras, programmable logic controllers, SCADA systems, and so on.

We saw how the iPhone allowed third parties to develop applications on their operating system; the same will soon happen with robots. For them, there is already a free and open-source platform called ROS (ros.org) which enables a worldwide, collaborative development of robotics.

Mechanical robots will soon populate houses, taking on the roles of personal assistants. This change will have unpredictable economic consequences that we will cover in the next chapter. The offshoring of jobs of all kinds — from the textile or automotive assembly industry to software programmers — will meet a new adversary: much lower cost machines, locally controlled and supervised.

The idea behind automation seems simple: Humans have limited strength and endurance. Exhaustion forces us to make elementary mistakes. In a training session, a professional player consecutively shoots dozens of triple shots without missing. But in a match, tired and bearing the pressure of the public and defenders, the situation changes. Stephen Curry, the best 3-point shooter of the last years, has not reached even 50% accuracy in any season. In contrast, robots guarantee greater process repeatability along with fewer failures and breaks.

We are used to seeing this idea applied to dangerous or burdensome tasks: application of chemicals and paints, assembly or loading of heavy parts, welding, etc. But automation will take an increasingly less physical, more intellectual drift. To understand this concept, we must begin by talking about two software technologies that will play a fundamental role. As they are software, they are not limited to manufacturing, but concern almost all organizations, no matter their activity.

The first software technology is robotic process automation, or RPA. RPA is not about mechanical robots, but imitating the typical tasks that a human performs on his computer, which would typically involve interacting with the graphical interface of a web or desktop application. In other words, we are dealing with a technology that learns what workers operate on their office machine and then executes them on its own. Despite the fact that I used the word "learn," with RPA we are not talking about anything like an artificial intelligence system yet. It works with simple, pre-programmed rules — RPA is cognitively stupid. It is designed to automate routine tasks that do not require smart decisions. Thus, humans can commit their creative brains at the service of the organization, instead of engaging in boring, repetitive jobs. No more cutting and pasting values in Excel or writing, again and again, the same data to the same interface fields. Among the most commonly automated tedious tasks, we find crossing and merging information, data entry, reporting, and monitoring updates to web page data.

RPA establishes the basis for a data culture and the cognitive applications that we will examine in a moment. The robot records everything. If an organization is considering adding artificial intelligence to its operation — particularly in real-time, such as a chatbot —, previously they must have automated most of the tasks involved.

It is possible to set up RPA by coding. This is the method employed by subjects with technical knowledge, where they configure the commands and tasks with source code. A significant advantage of this technology is that it simplifies the process and democratizes the technology to people without a specialized expertise by enabling graphical user interfaces and configuration with options menus. With these interfaces, the task is easily described using UML language, as in the example in the image. This is a very similar approach to BPM software. Another approach is to spy on what is happening on the monitor, a technique known as screen scraping: we just press the record button and execute what we want the robot to replicate.

The ability to exploit data is the most striking feature of this technology. Like industrial robots, RPA generates a tremendous amount of information and statistics about the operation. While RPA is a workable alternative to manual intervention, it should not be seen as a means to save money only. Too often, the question that managers come up with is: How many full-time employees will this robot spare us? It is the wrong question. Translating the RPA adoption argument into layoffs can lead to a dangerous spiral of productive deflation. The true potential of RPA should rather focus on the benefits it can bring to an operation: speed of execution, data quality, and integrity, employee satisfaction, and flexible work hours. RPA adoption also allows workers to serve on more interesting tasks, including interconnecting legacy systems or linking to third-party systems that cannot be reached in any other way.

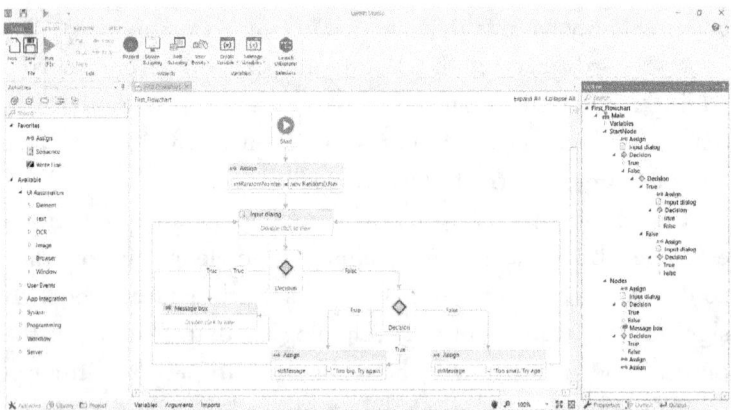

Figure 9: configuring a software robot from a graphic user interface. *Credit: UiPath.*

Understanding that automation does not pose a direct threat but rather a complement to the human workforce is also vital to avoid technological delusion. Software is no silver bullet. It is imperative to understand the tactical role it can play in strategic and cultural change within a corporation. Recall that RPA is the dumbest layer in the automation toolkit. The bricks. Keep this in mind: You are just removing the robot inside the human.

If RPA are the bricks, low-code development platforms are the concrete. We will use a simile to understand how they work.

A web page is not much different from more complex software applications: there is a source code that a compiler — in this case, the Internet browser — interprets, executes, and shows on the screen. In the past, web pages only displayed information, admitting no interaction with the user except browsing through hyperlinks. The pages were purely informative. There were some images and text. Over time, web pages became interactive entities, giving form to the so-called Internet 2.0. We then had pages that allowed a greater range of actions, like posting messages on our own. No one would pick up that text and manually insert it into their page. Since then, we have grown to build more sophisticated and robust web applications, of which the website is just the visible layer of something very complex. Unlike the closed software interfaces that we run on our local machines, we can consult the code of this outer layer. Right-click on any page and select the option "view source code" to review its HTML code. HTML is the language that browsers understand and execute to display a certain web page. Some browsers even allow you to go further. With the "inspect" option you can view the CSS code, the JavaScript events it is listening to, and much more. Today it is still possible to examine the code of any web page and learn to program websites in this way, but development technologies have become so difficult to grasp that it is preferable to take an online course.

When I wrote my first website over two decades ago, none of these programs existed; the CSS language was not there. CSS allows us to break up the data structure — the scaffolding on a website — from its aesthetic appearance: colors, fonts, etc. Long time ago, display style and information went together in the same HTML code. At that point, editors were rare, too.

What I did was consult dozens of websites and check their source code, one by one, testing in my browser to understand what each tag was for. I discovered, for example, that typing <p> gave input to a text paragraph and had to be closed with its corresponding </p> tag. The <img> tag allowed you to insert a picture, followed by a variable "src" to tell where that image was hosted. So, <img src = "/hello.jpg"> placed an image named "hello.jpg" on the screen. I found that when you called index.html, the browser understood by default that it was the home page, and so on.

Over the years, editors like Dreamweaver and Frontpage appeared, making website programming enormously easier. Nowadays there are websites like Shopify, Wix, or Webflow that allow us to make a web page in the same way that we create a Powerpoint slide: copying and pasting onto a blank canvas in a very intuitive and simple graphical interface. Today it is unnecessary to know about HTML, CSS, or Javascript or any other language to have our online website ready in a few hours. This does not erase the need for web developers, whose programming command allows them to produce professional, on-demand designs impossible to attain using an editor. But unlike what happened in the 90s, democratization has come to the web, and anyone with minimal computer knowledge can build their website with no help.

We just stated that web pages are the visible shell of something more intricate and profound. Well, this same democratization phenomenon is now replicated for all those other internal components. The same circumstance we are witnessing with web pages is being experienced in other pieces of software. Soon, we will all code software without knowing how to. Well, maybe not *all*, but many more people than in the past.

What are those other internal components? Let us look at a common architectural pattern called the Model-View-

Controller (MVC). In MVC, we detach data and business logic and write them independently, and call it the model. Its visual representation is called the view, and the module in charge of managing events and communications, the controller. All are written separately, like we do in web pages with CSS, but are more sophisticated. Think of any software you use. It contains millions of lines of code, so keeping them ordered corresponding to their function and separated into "boxes" has its advantages. This pattern allows the reuse of code and facilitates its subsequent maintenance. Today, when we buy a plane ticket on the website of an airline, we are seeing the "view" of software. By clicking and entering data, we are also causing events and calls to the "controller," who understands and analyzes them. For instance., interactions with the round-trip date forms. The software must then carry these tasks out according to rules, coded in the "model."

The great news is that in the same way that we came up with programs to make writing a website easy—a web which can be both a simple personal site or the "view" of a more complex application—more software is being developed to allow coding the other two layers (the controller and the model) in a much more straightforward manner. We can create full mobile or desktop applications without knowing how to code, or with minimal programming intervention. Not surprisingly, this new software family has been dubbed low-code development platforms. The amount of code required in low-code development platforms is little or nil.

RPA provides a sturdy structure to build on; it contributes to making an increasingly tall and slender building. Without sound foundations, a house would be too exposed to the wind or succumb to an earthquake. On the other hand, low-code development not only serves to keep it standing but also to make it more robust, it is as well the binder that elegantly allows us to go from one room to another. We can likewise look at RPA as a specific frontier of the low-code trend: software that requires little or no coding.

Artificial Intelligence is the culmination.

HUMAN, ALL TOO HUMAN

Kasparov stands up, furious, grunting, fussing with his hands and shaking his head, his left hand still gripping a pen. His eyes peep at his mother, who is in the room witnessing everything. He had just given up in Game 6 against Deep Blue. We are in New York, May 1997.

Although this contest was a front-page in all the newspapers, Kasparov had previously lost against other machines such as Deep Blue a year earlier. In the first game of the 1997 series, Kasparov won without major complications, but in this second, he resigned in a potential draw position, as shown in subsequent analyzes. It was this event that completely destabilized the world champion.

Why did a simple match make such an impact on the best chess player of the moment? The strength of the machines was in tactical chess—this was known to Kasparov. Algorithms did not play strategically. In other words, algorithms favored material gains over positional advantages, something typical of novice players. These tend to take pieces while neglecting other fundamental aspects of the game, such as structure: doubled pawns, weak pieces in the first rows, and so on.

Then Deep Blue arrived.

In that second match, Kasparov had set a trap for Deep Blue to win a pawn—material advantage—but lose globally on its position. Let us remember that no machine played with what the Great Masters call strategic foresight. Deep Blue had just glimpsed that. Instead of capturing the exposed pawn, Deep Blue chose positional advantage. In the subsequent press conference, Kasparov accused the machine of being "too human," implying that some Grand Master could be behind that or every movement. Deep Blue ended up winning the series, and in the sixth and final game, it brutally made Kasparov withdrawn in just 19 moves.

The move was 36.axb5![1] One of IBM's engineers, Murray Campbell, revealed a few years later that the action resulted from a coding mistake.[2] Many resources on the internet recount this moment. Whether or not that move was a mistake, the improvement of machines towards ever-increasing levels of cognition is undeniable.

In 2017, a Google machine defeated the champion of Go, a board game originated in China over 2,500 years ago that is similar to chess but considered much more strategic and complex to dominate by calculation alone. In comparison, Go is much more positional and less tactical. In 2011, fourteen years after Deep Blue, IBM once again amazed the world by defeating the greatest Jeopardy! champion, a television show in which contestants must answer clue-based questions that sometimes include double meanings. The first thing the machine needed was, of course, to understand natural language. The IBM chip beat the winner of the largest number of Jeopardy! shows of all time (Ken Jennings) and the player who had won the biggest award (Brad Rutter). Both were defeated in the same event.

That winning project's name was Watson. Since then, Watson has learnt new tricks, apart from playing games. Now it is on the market and accessible to anyone who can afford it. The project leader, David Ferucci, had a dental problem that caused him serious pain. He visited several dentists and underwent unnecessary endodontics. Finally, they referred him to other specialist areas, where they solved his issue. So, it occurred to him to design for Watson the first use case that has made him world-famous: medical diagnosis. Today, Watson is used by organizations in all industries: from virtual assistants (Watson Bots and Watson Assistant) to cognitive applications with data ingestion coordinated by the company itself (Watson Discovery). With some knowledge about the suite, it is possible to construct chatbots that suggest something to eat while explaining the recipe step by step, or an assistant who recommends an interesting movie—emulating the YouTube or Netflix recommendation algorithms. And IBM is not the only company dedicated to the development of cognitive systems that institutions of all kinds can exploit. AI is already making critical decisions across all industries and inferring behavior patterns from millions of customers.

With all this background, it is tempting to ask: is artificial intelligence intelligent? Howard Gardner, a Harvard researcher, first put forth his theory of multiple intelligences in 1983. Gardner tells us that, just as there are distinct problems to deal with, there are different human bits of intelligence to solve them. In particular, Gardner began by enunciating eight classes, expanded over the years to twelve: linguistic-verbal, logical-mathematical, visual-spatial, musical-rhythmic, bodily-kinesthetic, intrapersonal (self-awareness), interpersonal (ability to relate), naturalistic, emotional, existential, creative and collaborative. According to Gardner, a high-level athlete (say a soccer player), can have both a very high bodily-kinesthetic intelligence and low linguistic-verbal intelligence. This type of player will have awareness of his place on the field in regard to his peers, an innate talent to understand what will happen next according to the position of the ball and the movement of each player (what the sports press usually calls "reading the game"), and an excellent capacity to calculate the strength and angle necessary to move the ball to a specific point. At the same time, this player may lack competence in expressing himself off the playfield.

A corollary of Gardner's theory is that measurements like IQ just don't work. Intelligence is not a single quantity that can be measured with a number—at least not with a single one. Traditional intelligence tests for young students are usually focused on the first two bits of intelligence, leaving all others aside, becoming a source of frustrations for aspiring painters or dancers.

It is worth demystifying artificial intelligence. There is no effective intellect in machines, at least in the sense that we humans give. There is a herculean capacity for computation in machines, applied for certain specific activities that require it. Other bits of intelligence, such as the intrapersonal or the emotional, are still non-existent in machines, and may always be.

In the academic world, the term Artificial General Intelligence—AGI, sometimes called strong AI—is often used to refer to systems capable of imitating human intelligence. In particular, the expression is used to describe the ability of machines to generate abstract concepts from limited experience and, in the way humans do, forge relationships between different bits of intelligence to draw conclusions. For this reason, we call this artificial intelligence "strong," as opposed to "weak" or "narrow."

With the exception of science fiction movies, there are no strong artificial intelligence systems. Weak artificial intelligence systems, however, defeat humans in specific tasks often, especially tasks requiring high intellectual demand, such as games of chess. Analytical models have a strong statistical base; they cannot extrapolate that knowledge to other jobs. This is important to understand, not only for the limited applications that AI can have today but also because of the way these machines learn—their learning diverges from that of human beings or trained animals in many ways. What happens when you want to teach a computer to do something, but you are not sure how to do it? What if the problem is so complex that it is impossible to code all the rules and knowledge in advance? Machine learning allows algorithms to learn without being explicitly programmed.

Let us first distinguish between two types of learning.

Supervised learning occurs when a computer receives tagged training data, which consists of paired inputs and outputs. For example, we want a system that automatically searches for cat videos on Facebook, and delivers them every morning while we eat breakfast. We need to first make the computer understand what a cat is. For that purpose, we can input some images labeled as "cat," then more photos of similar furry quadrupeds, such as panthers, dogs, tigers, lions, and rats, appropriately labeled as "not cats." This instance is where another variant can come in, called reinforcement learning. We provide the machine with signals to let it know whether it is on the right track at detecting cats. Machines, however, are still far from humans in this regard. For a young child, building a mental construct of what a clown is will not be difficult once she has seen one at a party. If he changes clothes, wig, or makeup, she still understands that this is a clown. Instead, a machine will find it puzzling to see pictures of clowns from several angles: first in two dimensions, then through a 3D vision system. It will not understand how a clown can have sharp fangs like those of a feline, without recognizing that it is a fictional character created by Stephen King. It may seem inconceivable to use the term "clown" as an insult, and so on. These nuances differentiate human intelligence or AGI from current machine learning systems.

Unsupervised learning occurs when computers receive unstructured data rather than tagged and try to find inherent structures and patterns unknown to ourselves. Here, the machine teaches us. There are several frequent methods, such as association or clusterization. From heterogeneous data, the algorithm discovers common traces and creates groups.

Another fashionable term, deep learning, refers to the use of software that emulates layered neuronal networks. Deep learning involves mathematical structures inspired by the functioning of the human brain trying to solve problems. It is, therefore, a subtype of machine learning, which makes use of specific tools. In practice, we can utilize AI with simpler techniques and achieve faster and better results. Instead of worrying about creating personalized deep learning solutions, many companies opt for the commercial, almost out-of-the-box solutions of some startups or the most popular ones from Google, Amazon, IBM, or Microsoft.

Stanford developed an X-ray recognition software for chest complications, using a neuronal network system with supervised learning. They took an extensive set of x-rays and labeled them with the corresponding clinical picture: pneumonia, tuberculosis, or tumors—about a hundred thousand images. With this information, they trained the supervised learning system. Then, they compared it against human control groups. In an experiment with over 200,000 chest x-rays, the algorithm outperformed Stanford radiologists in diagnosing pneumonia.[3] We estimate that since 2016, algorithms read and interpret radiographs better than humans.

Thanks to the combination with other technologies, such as big data and the internet of things, the number of tasks performed by artificial intelligence are expanding, giving us more and more possibilities. For medical purposes, for example, we might have:

- Systems that detect, such as a heart rate meter connected to our phone or smartwatch.
- Systems that act accordingly, making no decisions other than following specific rules. For example, a system that sends an alarm every time the heart rate exceeds 120.

110 | A robot took my job away

- Systems that predict. One that understands that we go out to run every morning at eight o'clock, so a heart rate surge then is perfectly natural. Thus, it does not ring the alarm. Instead, a raise at five in the afternoon will not be considered normal.
- Systems that learn and teach. For example, a system that learns that on Saturdays we force our beats more than necessary and then suggests that we rest that day, instead of Mondays.

Machine learning systems are now capable of composing Beatles-like songs,[4] painting pictures and exhibiting (several communities of human artists make art through AI, such as aiartists.org), or writing newspaper articles. They can even invent humans that do not exist. See figure 10.

Figure 10: none of these people exist. They have been created by a computer. *Credit: HackRead*.

Natural language processing (NLP) is starting to be implemented in several call centers. Machines that communicate with us in Spanish, English, Russian, Japanese can also recognize if we are angry and, in that case, give way to a human operator.[5]

What we talk about when we talk about innovation

Despite all this, cognitive systems are still poorly advanced. They need to see tens of thousands of clowns before they can understand what a clown looks like—but they have some advantages over us. For example, the systems don't forget. We know we humans have a limited capacity to absorb information, in terms of time and quantity. We forget topics we don't practice, languages we don't use. The Hebbian theory tells us that the value of a synaptic connection increases if the neurons on both sides fire repeatedly. Based on this, there exist many learning methods. Quizlet or Memrise, for instance, help us learn vocabulary by repeating certain words from time to time, making them stick longer in our memory. Machines don't need this. As we saw in the first chapters, digitized information is not prone to be lost due to the degradation of the media, unless we destroy it. We can always expand disk storage or RAM space to a computer. Humans cannot, *for now*.

In the same way, machines will soon overcome some advantages we still have over them. The captcha used in some websites to check whether we are human will disappear. They are based on a kind of brain plasticity that only humans have, to recognize crooked letters, with different fonts, in bold, in italics, even mixed with numbers. But that soon will be replicated. Can you read this text?

> *Aocdcrnig to rseecrah at Cmabrigde Uinervtisy, it dseno't mttaer in waht oderr the lterets in a wrod are, the olny irpoamtnt tihng is taht the frsit and lsat ltteer be in the rhgit pclae. The rset can be a taotl msses and you can sitll raed it whoutit a pboerlm.*

This text is inspired by Graham Rawlinson's doctoral thesis in 1976 for the University of Nottingham. Brain plasticity, however, has its limits. The brain, for example, finds it difficult to read longer words and feel exhausted just by reading sentences above twenty words. We study this in most literary style manuals; it is a well-known phenomenon. When trying to mess up the letters in more complex sentences, the result is also much more complex.

> The adkmgowenlcent—whcih cmeos in a reropt of new mcie etpnremxeis taht ddin't iotdncure scuh mantiotus—isn't thelcclnaiy a rtoatriecn of tiher eearlir fidginns, but it geos a lnog way to shnwiog taht the aalrm blels suhold plarobby neevr hvae been sdnuoed in the fsrit plcae.

The original text is:

> The acknowledgment—which comes in a report of new mice experiments that didn't introduce such mutations—isn't technically a retraction of their earlier findings, but it goes a long way to showing that the alarm bells should probably never have been sounded in the first place.

Machines don't have brains. Neuroplasticity is imitable and trainable by software—there will come a time when algorithms can decipher captcha in an equal or better way than humans do, and will even be able to interpret the writing of doctors. There will come a time when we will not be able to identify what is real and what is not. The algorithms are already writing a colossal amount of the published content on the internet.[6]

Thanks to the development of NLP technologies, the machines will assist us on behalf of organizations that today maintain human-based call centers. The tiny chatbots that we can see on some web pages, which still answer rather simple questions, will expand to more and more tough use cases. Thanks to NLP, they will answer directly in our language.

This is just the beginning.

DIGITAL METAMORPHOSIS

4

> *"My interest is in the future because I am going to spend the rest of my life there."*
> —Charles Kettering, 1876–1958

In 2005, the Rails framework became notorious after its authors advertised in a video being able to code a blog engine in less than 15 minutes.[1] To accomplish such a feat, they used a little trick: a scaffold of self-created code they summoned using previously programmed commands. This is precisely what a software framework is: a library that saves work when programming the most common functions (create, edit, and delete users; implement login and log out; assign permissions, and so on). Instead of coding everything from scratch, you type in instructions that describe the desired behavior. A pre-defined script generates the new code—*the code writes itself*.

Ruby on Rails's creators understood that they repeated tasks between distinct pieces of software, so they saved work by pre-programming them. Similarly, a large percentage of innovation today is about arranging typical pieces in ways that had not been used before to meet unmet needs. For example, a few years ago, there were no apps to request a taxi;

now there are many. All these applications comprise a login system, which most times is done through social networks (social login), a payment gateway to register and pay with our credit or debit card, external systems such as PayPal, a GPS, an algorithm for deciding the best route between two points, maps, etc. Let us think about Uber. The assembled package did not exist, but most of its parts did; the pieces were there. This has allowed the appearance of frameworks such as Rails, or software families such as RPA and low-code platforms. They reuse components from one application to another and allow "programming without programming." We tell the machine what we want to do and then it writes the corresponding code. This ensures that not only those with advanced programming knowledge can access the power of the "already-written code universe."

Consequently, the global development pace has sped up rapidly. This increase in speed affects how often new applications hit the market, and with them our daily behavior changes. In 2014, the consulting firm Boston Consulting Group updated its famous growth matrix, originally published in 1970 and studied in all business schools in the world. The matrix describes a 2x2 grid where the distinct types of business or economic activities of a firm are characterized. On the vertical axis is the annual business growth, while on the horizontal is the market share of the company. Each business unit or product will land in one of the four quadrants according to its strategic value: stars, with great growth and market share; questions, growing but with little share share; cows, low growth but high share—profitable but declining products which need to be milked; and dogs, without growth or participation, that need to be terminated.

In 2014, BCG declared that the matrix was still in force but needed some clarification:[2]

What we talk about when we talk about innovation | 115

> *First, companies face circumstances that change more rapidly and unpredictably than ever before because of technological advances and other factors. As a result, companies need to constantly renew their advantage, increasing the speed at which they shift resources among products and business units. Second, market share is no longer a direct predictor of sustained performance. In addition to share, we now see new drivers of competitive advantage, such as the ability to adapt to changing circumstances or to shape them.*

According to BCG, the matrix has not lost its value. Business flows, however, need to be applied with greater speed and focus on strategic experimentation. The matrix also requires a renewed horizontal axis, as the current market share is not a sufficient predictor of future performance. Why? Disruptors can steal it overnight. Entry barriers to the digital world have dropped significantly. We require more capacity for trial and error.

Those who think their market share is untouchable because of regulatory restrictions or high access costs are in for a surprise. They must ensure that they will have the ability to quickly respond to their competitor's threats. Those in need of transforming their operations are easy to identify. They all display the same traits: outdated and isolated computer systems, which do not capture key data to make available to its users; a short-term vision, which focuses on meeting current objectives and penalizes investment in future innovation; an inability to attract new young talent; a prevalence of ancient employees; undocumented, obsolete, and manual processes; work in silos; and strong organizational politics, especially when allocating and prioritizing resources.

Outdated institutions are unprepared for today's pace because they operate under a mistake minimization model, inherited from Taylor, Ford, and the factory manufacturing from the last century. The speed at which we currently innovate, (the increasing importance and ease of access to software) put many companies at risk, even if their activity has nothing to do with software. And not just them—it affects the entire economy.

Barriers have been broken. Certain technologies allow what a few years ago was impossible. And it has been democratized. People who desired to form a business needed to have a product, a channel, and to broadcast to attract attention. Centuries ago, each of these three aspects constituted an insurmountable handicap for those who did not count on an enormous capital. To produce on a large scale was reserved for the wealthy only. Most times, it still is, but nowadays we can outsource or manufacture an endless cast of services over the Internet. Digital products increasingly replace physical ones. The music trade nightmare came from someone who did not have an expensive factory but could distribute songs almost freely, remotely, only for the cost of the connection, and without having to buy blank tapes. Today, software like Shopify allows having our online sales channel, and social networks, digital marketing, or YouTube, to spread it. Now, it is more likely to learn about a product through the Internet than through a television commercial. ECommerce specialists reach our social networks with customized ads, breaking with the traditional advertising industry.

The business world has changed and with it everything else. Moore's Law reminds us that the size of processors is halved every two years. Airbnb had less than a million bookings in June 2010. Bookings rose from to two million in June 2011 to five million on January 26, 2012 and to ten on June 19, 2012. If we draw a line with these data points, we will observe an exponential growth graph. The opposite case, exponential decrease, is found in the time it took for businesses to reach a billion dollars of market capitalization (an average Fortune 500 company would take approximately twenty years). Google, in 1998, took eight years and Facebook, in 2004, about five. Tesla took four while Uber and WhatsApp, which went public in 2009, took just over two years. The term unicorn refers to startups that take a very high market valuation in a brief space—so little so they don't have the time to lose their startup status. A unicorn is thus a startup valued at more than a billion dollars. The term was coined in 2013 by venture capitalist Aileen Lee, choosing the legendary horse to represent the statistical rarity of such firms. It is not so that difficult to see them, however: there were 279 unicorns in March 2018. Some examples are DiDi, Airbnb, Lyft, Stripe, or Palantir Technologies.

This exponential pace will generate headaches at several levels. The first headache is educational. Humankind has a linear ability to adapt and learn. We do not seem to be able to cope with the current velocity of moral and political challenges or absorb the amount of information that technology is proposing. The second headache is suffered by the workforce. Organizations are experiencing a similar phenomenon of linear adaptation versus exponential reality known as Martec's law. They cannot adapt to such a tempo. It has happened in various industries and will happen in many others that, as I write these lines, do not even suspect it. The third problem is at the productive-system level. Until now, we have seen this exponential paradox applied to intangible entities such as software and data, but some new emerging technologies can cross the borders into the physical world. A greater ecological challenge comes with higher production capacity.

Let us analyze these three questions in more detail.

THE TRANSFORMATION OF EDUCATION

Everything is pointing towards a substantial mutation in the way we learn and educate ourselves, and it looks like it will happen during this decade.

In 2011, two Stanford professors named Sebastian Thrun and Peter Norvig posted a series of videos on YouTube that took advantage of the platform's new functionality — actionable buttons on the screen — to link chapters and introduce short questions and quizzes. The course was free, advertised by Stanford. Everything was online, including the assignment of grades and a certificate issued by the academy. It was crazy. 160 thousand students registered; two-thirds were outside the United States, spread across 190 countries. In Iran, where YouTube was blocked, a student copied the class website and released the videos — with the teachers' permission — to one thousand other pupils. Although earlier massive open online courses (MOOC) were being offered by the University of Manitoba (Canada) since 2008, Thrun and Norvig's course was the first to have a significant impact worldwide.

We can draw some very interesting conclusions from the course, which lasted ten weeks. The first was that there was a niche market for online training provided by eminences in the field or prestigious institutions like Thrun, Norvig, and Stanford itself. Those same professors founded Udacity soon after. They were rapidly joined by Coursera — with an initial investment of $22 million and a constellation of agreements with universities —, edX — from Harvard and MIT, and a seed capital of $60 million —, and their British counterpart version, FutureLearn, operating from the London neighborhood of Camden Town. The fever went on until the bubble was somewhat punctured in 2013, with a study[3] that proved the course involvement and completion levels were rather low. Far from premature death, this resembled the typical errors of emerging technologies. MOOCs, although not having remotely replaced any traditional institution, are positioning themselves better each day to partially oust them. The exact percentage in which current higher education will be replaced over the coming decades is unclear. Some claim that they will be erased from the map, which I consider unlikely. Others think that MOOCs will complement them, making learning times and spaces more flexible. Thrun sees a future with up to ten institutions offering higher education, which fits well with the classical "winner-takes-it-all" scheme of disruptive digitization, such as the case of social networks, where even Google was not able to compete.

There was a demand for online courses, but further conclusions were drawn. The data revealed something deeper. Among all those enrolled in the course, Stanford's top-ranked graduate finished the course graded below the 400th place. The program was open to everyone. Everybody, Stanford and outsiders, received the same training and the same scoring system with the difference that none of the outsiders had paid the $40,000 a year that Stanford's tuition costs—student debt is a recurring concern in the United States. The consequent question here is: Can private and elite institutions afford to participate in the MOOC industry, with relatively short and cheap courses? The a priori answer is a resounding no. Private higher education is not as deep in the teaching business as it is in the accreditation business. Their diplomas work in a very similar way to paper money: if many bills are printed, they generate inflation and lose value. If those institutions choose to enter this game, it will have to be on tiptoe while measuring each step.

Before classes would end, Stanford management knew about the magnitude of what was going on. They called a meeting to decide what kind of accreditation they were going to give to the students, who were already promised something during the registration process. A part of the concern was legitimate: in this incipient state of affairs, there were no mechanisms to check if students had cheated in their courses —biometrics, particularly facial, will play an important role in the future. The remaining anxiety directly involved allowing tens of thousands to move around the world with a Stanford diploma under their arms. It was agreed to send a note to the scholars who finished the course, avoiding any official document. When a journalist used the word *certificate*, Stanford quickly requested a correction in his press release.[4]

Those interested in the future of education should carefully follow the evolution of the format. Regardless of what and how much are they going to change, MOOCs are an excellent opportunity for firms to offer quality training to their employees. Big corporations like AT&T, General Electric, or L'Oréal have signed collaboration agreements with MOOCs and the trend continues. There are multiple advantages and benefits to these collaborations. Workers can access anytime, anywhere, rather than having to wait for a scheduled session. It ensures that employees are up-to-date with the latest knowledge, an aspect that organizations that produce their own material sometimes neglect. And even for those who want to stay in control, the syllabus and external appearance of MOOCs can be customized. Members can indeed participate in branching scenarios to find the education path that best suits their daily tasks, instead of massifying and homogenizing the same content for everyone.

Outside of the academic world (but related to the way we acquire information) we observe a change in content consumption behavior. We have experienced in the past decade a shift from conventional media to new forms of presentation, particularly social media and online news sites. The landing of YouTubers and influencers replacing old radio and television stars is just another symptom of the same medical picture: information flow has become democratized and bidirectional. This has its advantages and disadvantages; some point out the risk of having individuals not formally educated in journalism acting as such. But these voices ignore that guild intrusion is ancient, and trained subjects have not prevented the previous manipulation from the media.

In 1988, the linguist Noam Chomsky published his classic Manufacturing Consent, which discussed the independence of the press and the most widely used forms of manipulation of the public opinion. A few years later, he offered a well-known BBC interview with journalist Andrew Marr. Marr sarcastically asked if the press was self-censoring for not publishing certain things, to which Chomsky replied: "Not self-censoring. There's a filtering system that starts in kindergarten and goes all the way through and it doesn't work a hundred percent, but it's pretty effective. It selects for obedience and subordination (…) I'm sure you believe everything you're saying, but what I'm saying is, if you believed something different, you wouldn't be sitting where you're sitting [doing the interview]."[5]

Digitization does not seem to be allowing an entrance to a novel, greater or unprecedented information manipulation because it already existed. But it is true that topics that seemed overcome reappear with vigor, popularity, and attention. Everything is exponential, the good and the bad. Anti-vaccine movements, flat-Earthers, influencers prescribing antibiotics without knowing what they are doing[6] —all of this sounds troubling, but it is more anecdotal than anything else. No major regulation has been changed in this regard. The interesting thing will be to monitor, in line with Chomsky's statement, how online education influences our behavior as a society in a few decades. Will freedom of information make us free of thought?

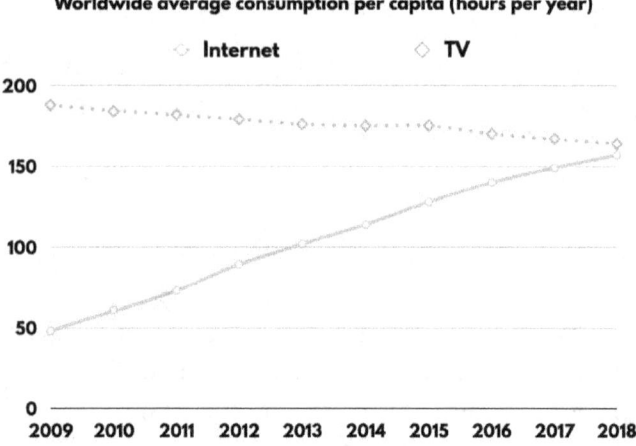

Figure 11: Average television and Internet consumption per capita worldwide. *Source: Statista and Business Insider*.

THE TRANSFORMATION OF LABOR

It is an ancient phenomenon, but in the second half of the last century, developed nations intensified the offshoring of some of their manufacturing activities towards developing countries. It is a simple event to understand: wages in those regions are lower, so it is possible to dramatically reduce production costs. Despite union pressure, entire industries such as the textile or the automotive industry moved most of their capacity to foreign countries. The same thing happened with software development, which was mainly transferred to India in the 90s. Between 2004 and 2015, jobs were relocated at a rate of between 150,000 and 300,000 per year in the United States.[7] Surveys indicate that between 76% and 95% of US inhabitants believe that manufacturing relocation is the reason why the economy and local employment are suffering.[8]

The three trends aforementioned—the rising importance of data, platforms with low entry capital, and ubiquitous connectivity—affect this phenomenon. The end result is uncertain, but some forces will act as a counterweight to offshoring. First-world nations will cut production costs and more skilled workers will be needed in exchange for less cheap labor. Offshoring activities will be brought back, transformed into near or local services. Digitization will open the doors to a manufacturing resurgence in rich countries, replacing manual capacity with software and automatic processes. The added value of open innovation with leading universities and research centers headquartered in developed regions will be amplified. Any activity prone to be delocalized today might be automated and repatriated tomorrow—even those that require human intervention. For example, we have been working on the automation of call centers for some years, starting with some basic use cases—requesting the status of an order, checking the account balance. Increasingly, relocated call centers will be replaced by robot voices.

The impact of digitization in the labor balance sheet is already observable. Computer technology employment recovered in 2015 to pre-2001 figures in the United States. Yes, other sectors still continue to relocate at an unstoppable rate, but in total summation, reshoring (the process of returning the production and manufacture of goods to the original country) reached record levels in 2018 for the United States,[9] recovering some 800,000 manufacturing jobs in the decade spanning from 2010 to 2019.

Even without repatriating to the original country, an intermediate phenomenon called nearshoring can be seen. Nearshoring consists of bringing back jobs to a geographically closer area. For example, in the case of the United States, a company would move a production plant to Mexico or Costa Rica. It seems that nearshoring will experience considerable growth in the coming years, spurred by greater time-zone compatibility and geographical and cultural proximity. This represents enormous opportunities for economies adjacent to the most industrialized nations. Take, for instance, a software development lab situated in a border zone between Mexico and the United States, versus the classic relocation of this type of activity to the Asian continent. The first situation would generate indirect jobs for both countries, such as hotels, transportation, and infrastructure, as well as taxes that would be paid by the company and by employees who have local contracts. When the operations are located in another country, all of this is lost. In the case of tangible goods, a large number of jobs would be gained in the transport, logistics, distribution, and retail sectors. The paybacks are distributed more fairly and equitably, too: offshoring tends to favor the capitalist partners over the workers because the labor and tax benefits of the work carried out offshore rest mainly with the former. But when jobs come back, everyone is better off.

It also turns out that offshoring is not sustainable *ad eternum*. Some target regions—namely the smallest societies which had free trade zones and a government that collaborated with heavy investment—developed enormously since the end of the last century and adopted the advanced technical knowledge required to produce and export technology. It is the pioneering example of the Asian tigers. These once unindustrialized nations, South Korea in particular, compete today in numerous industries against the most advanced countries. Their costs do not differ greatly from those of the First World—some even exceed them. The Asian tigers are tiny countries of minor demographic importance, but India is developing a major national auto industry and for years, China has led the mobile phone market.

In December 2017, Apple CEO Tim Cook visited China as a guest of the Fortune Forum. On the iPhone cases, it reads:

"Designed in California, assembled in China," so he was asked if this was due to labor costs. Cook replied:[10]

> *There's a confusion about China. The popular conception is that companies come to China because of low labor cost. I'm not sure what part of China they go to but the truth is China stopped being the low labor cost country many years ago. And that is not the reason to come to China from a supply point of view. The reason is because of the skill, and the quantity of skill in one location and the type of skill it is. The products we do require really advanced tooling, and the precision that you have to have, the tooling and working with the materials that we do are state of the art. And the tooling skill is very deep here. In the US, you could have a meeting of tooling engineers and I'm not sure we could fill the room. In China, you could fill multiple football fields.*

This fits the circle for developed nations: on the one hand, they have the opportunity to drastically reduce their costs and, thanks to these savings, afford more expensive labor (their own), which in turn entails logistical, cultural, and hourly facilities. On the other hand, they begin to feel on their back the threat of rising costs in some emerging states.

Something relevant and often not considered is that, despite the rapid growth of the Chinese industry, the United States maintains a comparable annual total manufacturing production,[11] with a population equivalent to a quarter of the Chinese. This reflects that developed countries remain highly productive. Therefore, competitiveness is a matter of salary cost. There is a threat that phenomena similar to those described by Cook in China for advanced tools will be replicated for other regions and industries, disrupting the model of offshoring companies.

It is true that Cook was in the Chinese city of Guangzhou and that, in front of the public, it is not easy to make a speech stating: "we come to China because it is cheap." His argument cannot be applied to all industries, nor to all Chinese workers. But what he describes is not new: places such as South Korea or Taiwan went through the same cost transformation. The same phenomenon might be replicated in other regions and sectors. Now are the MINT (Mexico, Indonesia, Nigeria, and Turkey) the candidates to develop in the forthcoming years. Previously, it was the BRICs (Brazil, Russia, India, and China). Nearshoring policies in Mexico for sectors such as the textile or automotive could collapse if Mexican growth expectations are met in the following decades. India produces even more cars than Mexico. Brazil, Thailand, and Turkey are in the top 15. In the textile sector, Bangladesh, Vietnam, and India pursue China and Germany as the top main producers. Turkey is seventh and the rest—Spain, the United States, and Italy—have high wage costs. With these data, what can be expected to happen in these two sectors? How will the relationship of productive forces between regions evolve?

In 1999, Robert Mundell was awarded the Nobel Prize in Economics. Mundell, despite being Canadian, is known as "the father of the euro" for his work on optimal currency areas (OCA). He said that for an OCA to exist, two conditions should be met: the convergence of macroeconomic factors and the freedom of movement in regard to capital and labor. When the Eurozone was created, macroeconomic figures indeed converged—Maastrich demanded that all the countries joining the euro should have inflation, total debt, and public deficits lower than certain values. After the 2008 crisis, Germany forced measures to combat public deficit and debt contrary to the Mediterranean countries' interests, all with high unemployment. While the formula to raise the aggregate demand and fight unemployment is to lower taxes and interest rates while increasing public spending, if you want to reduce deficits and debt, you need to do the opposite. The first of Mundell's conditions was crumbling. But then, after the crisis, more problems arose with the second factor: while capital fluctuates freely—even outside the Eurozone—labor does not.

Workers around the world must have work visas to migrate and, even in free transit areas such as Europe, it is not so easy to fulfill in practice. For a Sicilian, it is not as easy to move up to Finland to work as is to make an international bank transfer. Some Spaniards and Greeks moved to Germany, but the unemployment rate differences remained dismal. There was no offshoring between the Mediterranean and Northern Europe. But from now on, workers may not need to physically relocate to get the work factor moving around. Connectivity will be the new means of transportation.

How will the digitized labor reality affect goods (both digital and physical) in the future? It will depend on how countries capture global demand. Increasingly, the planet will resemble the Eurozone, with free flow of workers, except that they no longer need to move physically. Jobs will go where demanded by the overall economy. Europe and the United States have a chance. Strong exporting regions, such as China, Taiwan, Singapore, and South Korea, may be affected by the changing cost structure. As in the case of Europe, which also has large exporters with Germany or the Netherlands, everything will hinge on whether they know how to attract or retain talent. But for some developing nations that rely on imported manufacturing, this can become a giant opportunity, especially with open-source projects that allow the use of new technologies at low cost.

When the adoption of automation will be sufficient, certain offshoring jobs will reduce their marginal costs to almost zero. Some will be repatriated. The next things we will have are specialized tasks that cannot be delocalized, but can be directly automated. An example, simple but that we can already appreciate today, is paying in some supermarkets and retail stores that replace humans with machines, which allows customers to read barcodes and pay with minimal support. This is still little tech and a lot of rebalancing the workload from the company to the customer, but represents just a small clue of what awaits us. Many of the activities that we now consider irreplaceable by machines may soon be, albeit partially, due to artificial intelligence and remote control. A recent study hypothesizes that the demand for jobs that require a high level of preparation peaked around the year 2000 in the United States and, since then, has been in constant decline.[12]

THE TRANSFORMATION OF MANUFACTURING

The democratization of some technologies (especially 3D printing) will open the doors to social production, spurred by the growing trend towards personalization. We may soon design our own tools, clothes, and even our house or cars. Whether or not we will host domestic 3D printers anytime shortly, the truth is that companies will manufacture digitally in their places of origin, even directly at points of sale. The production chain will undergo a complete conversion. Any trade balance between China and other countries — the United States in particular — will be affected.

Digitization opened the doors to create service apps with small capital investments. New business models were born, such as dropshipping, leveraging platforms like Amazon, and selling online without the need to accumulate stock by taking an order and passing it to the wholesaler. But you always needed a product made by someone else. Pieces will soon be printed with a cost reduction similar to what we have experienced in other aspects of the trade. Economies of scale and mass production may no longer make so much sense. This will pose not only economic but legislative challenges. A printed Ikea table made from identical materials and digital design by another company may be illegal, but is it false?

Perhaps we would redefine our concept of piracy. Does the transfer of a digital design need to pay customs tax? Can this type of tax be legislated effectively? What already happened with songs and movies will happen again with tangible objects. Within companies, new labor relations, statutes, and laws will arise to cope with a world where remote work will be the norm, not the exception. Who will benefit from all this?

3D printing is so crucial because it opens up the possibility for physical objects to access the exponential spiral into which intangible things, such as music or video, previously entered. There are no restrictions on a digitized planet. You can make infinite copies of a song without affecting anyone or taking away his copy. The line that separates the digital from the physical will narrow more and more until it fades. Some manufacturing techniques will become obsolete, and in return, software more critical—if possible. 3D printing technology was born in the 80s and the industrial field has used it for more than two decades. Why hasn't it matured yet? Similar to the electric car, which requires the coordination of a large cast of technologies, 3D printing relies on high-precision heads, lasers, plastic adhesives, advanced extrusion, and the development of appropriate software and raw material preparation—for illustration, some techniques require previous processing to spray the material. Today there are 3D printers that use ceramic, aluminum, glass, and metals. Relatively cheap printers are already capable of printing with an accuracy of 0.1mm. Many procedures are still slow, but they continue to develop.

Throughout history, the methods humans have used to manufacture their objects fall into two families. The earliest and most primitive human tools consisted of taking a piece of material and cutting, filing, scraping, or sculpting it until it was properly shaped. Thus, this method involves removing material. Some of the greatest works of sculpture were made by carving on marble or stone. The second method is quite the opposite: adding material and shaping it, either with your own hands, as the clay artisans do, or by means of a mold, as many objects are still made today. For example, ship hulls made of fiberglass use a particular design. Both methods are problematic: in the first case, the efficiency of the raw material is very low, since the one that is discarded in many cases cannot be reused. We throw away huge amounts of valuable material. In the second case, the generation of the mold itself can become very laborious, as in the mentioned case of boats.

3D printing, whose technical name is additive manufacturing, is a very subtle evolution of the second method, except that it does not require the construction of any mold nor the application of subsequent force to sculpt the material. Using an additive method with 3D design is important because it considerably decreases the amount of waste during production. Compared to molding, where we make a mold for each part and certain operations are impracticable, our possibilities increase. A 3D printer obeys instructions from a program and is indifferent to the complexity of the object you are printing. With the same printer, we can print innumerable different objects. Thus, the door opens to a wide array of personalization options, something critical for many business models. 3D printing takes us back to the times before the Industrial Revolution when the massification due to the economy of scale broke with the previous forms of manufacturing, which were expensive but personalized. In the past, clothing was made to measure, at home, or commissioned to a tailor. The furniture was unique, custom-bought from a carpenter. The first Ford cars were constructed almost entirely by hand and came in various colors. After the introduction of his assembly line and the mass production that allowed Ford to reduce costs, there came his famous quote: "A customer can have a car painted any color he wants, as long as it is black."

Additive manufacturing will also contribute to the environment. It will reduce the enormous expense and pollution produced by the global logistics of assembly parts. The manufacturing hubs will be relocated and go local. We will transfer digital files instead. There will be less stock shrinkage, less waste, less economic loss, and just-in-time manufacturing will be radicalized. The qualities could be adjusted, because the same piece would be made of better or worse materials simply by recharging the printer, reusing the same blueprints. If humans continue to pollute the planet at current levels, it will be due to our consumption model, not the production. This technology is a candidate to alleviate many of the evils that surround and menace humankind.

Figure 12: This is the first printed habitable house in Europe, built in Russia in 2017 with a plan of 300 square meters. *Credit: Wikimedia Commons.*

Given the shortage of housing in some areas, the price bubbles, and our political incapacity to efficiently manage empty real estate parks, we would choose instead to buy cheaper land away from the cities, where the construction of a 3D printed house is already possible and where the most radical cost reduction is expected. In France, several corporations, including giants Saint Gobain and Vinci, are already building using XtreeE technology.[13] The D-Shape printer, one of the most popular in this field, was patented in 2005 by Enrico Dini and since then has been used to construct everything from bridges to sculptures.

With the advent of 3D printing, it is worth asking what role architects, interior designers, and other professionals, now using software like AutoCAD, will take on. Possibly, their adaptation to the world of 3D printing will involve adopting new programming skills.

In the health sector, we find 3D-printed prostheses, dental implants, and hearing aids. And in the coming years, on-demand drug design will advance fabulously. In 2015, Aprecia, a futurist pharmacy company exclusively focused on manufacturing printed medicines and developing the necessary technology to obtain them obtained the license to market Spritam. It is the first printed medicine available, printed with a proprietary technology called Zipdose.[14] The FDA itself released a press statement[15] in 2017 with a series of recommendations for manufacturers who are seeking to enter the commercialization of this class of medicaments. In parallel with pills, the development of printed human organs is being recklessly investigated. Science fiction? It is still so for complex organs, such as the liver or kidneys. But since 2004, Luke Massella is the first person alive with an implanted 3D-printed bladder. At the end of 2018, he was still surviving with the same manufactured organ.[16] Advances in organ printing, if they become a reality, will be particularly important in educating future generations of doctors and in improving the success rate of transplants. The immune rejection of third-party donated organs will be lessened because the new organ's impression is made with cells of the patient himself.[17] If you are interested in this topic, I highly recommend you visit the website all3dp.com.

The same bioprinting concept can be applied to food, although it will be difficult for us to get used to the idea. In a few decades, we will be eating printed meat, which will greatly improve the ecological aspect of our diet and lessen instances of current animal abuse. Remember that the pressure towards meat consumption grows every year due to the enormous inefficiency with which we produce it. It does not appear that most of the population wants to give up meat in the coming years. Printing meat is an ethical and ecological alternative to vegetarianism and it offers a wide range of additional benefits, like including substitutes to animal protein, such as algae or insects, in foods easier to consume. It also opens the door to food personalization and will facilitate a healthier, more measurable, and controllable diet by doctors. In some parts of the world, it is already possible today to taste printed food, such as some Starbucks coffee shops in Los Angeles that offer 3D printed ice creams from Dream Pops (see image).

Figure 13: 3D-printed ice cream produced by the Dream Pops company.
Credit: rarenorm.com

Will 3D printing reach homes soon? As with all technological adoptions, not yet. In some places, it will come earlier, in others later on. The relative impact of its domestic use is yet to be verified and stated in relation to the industrial and macroeconomic levels. For personal use, business models that allow digital files to be uploaded to the cloud and printed on demand in warehouses will become very popular. But on a global scale, 3D printing can become huge.

One thing is for sure: digitization will cross the border of the intangible to sneak into the physical world. And this will have outcomes still difficult to predict.

PRACTICE

HOW TO INNOVATE

5

*"The first principle is that you must not fool yourself—
and you are the easiest person to fool."*
—Richard Feynman, 1918–1988

In 2006, actor George Clooney starred in a Nespresso commercial for the first time. Dressed in an elegant black suit and living a life of luxury, the commercial was reminiscent of those produced by Martini a decade earlier—the long tradition of advertising fine tobacco, perfume, and expensive clothing was nothing new. In this case, however, Clooney—a sex icon—promoted something banal: coffee. And he did it for a relatively unknown brand, despite having been on the market for two decades and belonging to the largest food conglomerate in the world.

Milestones in Nespresso's history seem to occur every ten years. The original design of their coffee machine is from 1976. The initial proof of concept happened in 1986. In that same year, they established as a company within the Nestlé group and got their early significant sales in Japan and some countries in Europe. They filed their first patent in 1996. In that same decade, *Le Club Nespresso* was born, cementing the luxury brand that it would soon become.

Nespresso's secret is not in the coffee, but in everything that circles it: the coffee machine with hundreds of associated patents and above all, the glamor that they generate over the brand. The stores, the experience, the social status all contribute to their advertising strategy. *George.* Other manufacturers, such as Nescafé or Koblenz, quickly replicated the pod coffee maker system. Nespresso's sales have not stopped growing, however, despite the competition. They did not invent an original drink. Even turning coffee into a luxury product was not new. Starbucks's founders had already figured this concept in the 1970s after a trip to Europe. Starbucks wanted first of all to be a delicacy coffee shop.

One of the greatest myths about innovation and entrepreneurship is that you need a brilliant, unique idea that no one else has ever thought of, which is typically a device. *What* could we invent or sell? It is the most frequently asked question in companies. The *how* rarely comes to the scene. Sometimes the *who* or *where*. Never the *when* or the *why*.

The dangerous liaisons: obsessions with the product while simultaneously neglecting the customer. We shut ourselves up and ask inside our drawer: what is this thing that nobody sells that I could develop? The problem is not only limited to innovation but also affects current products in our portfolio. By ignoring the ramifications, we miss sales opportunities.

Something to keep in mind when evaluating a product is the difference between quality and utility. Some authors refer to both concepts using different words, so let us define them. By *utility*, I refer to the ability to satisfy our basic needs with a product. Does the product work or does it not? With *quality*, I speak of the wide range of functionalities and additional considerations that surround it that make it more suggestive

What we talk about when we talk about innovation | 139

for the consumer, even challenging reasoning. With quality, we suggest its allure, not its workmanship.

Let us take the case of a utility car and a Ferrari from the point of view of a user who needs to drive 10 kilometers to his job and back. If both cars can perform this action, meeting this need, the buyer will automatically go on to test other factors related to the quality, *according to his definition*. This is important because when we design products, our minds travel towards things that make sense to us, without worrying about whether they also have meaning for our target clientele. The utility is usually abundantly clear and simpler to perceive: we know what a fork, a sock, or a book are for. A cigar is used for smoking, but this act can be carried out to calm anxiety, submitting to addiction, or liking the boy next door. Deep down, the one who smokes to attract the guy does not enjoy tobacco—in this case, inserting a menthol capsule into the filter seems like a splendid idea. In other situations, that same capsule is a sacrilege for those who enjoy the original flavor of tobacco, and so on. The second derivative is terribly subjective. This is why agile discovery and design thinking techniques work. This is why customer empathy and segmentation are so powerful.

Sometimes we take for granted that utility is always fulfilled. This is not invariably the case. A telecommunications operator with poor network quality will be seriously affected by the allure of its product regardless of the brand, the quality of customer service, or the color palette they use in their advertisements. Your ultimate goal is connectivity, and if you're not able to provide it—say because there are too many shadows in rural areas and the signal fades away while driving—you don't provide a useful product. You may attempt to decorate your product by using attributes that are more characteristic of quality: an attractive price, advertising, continuous promotions, etc. But then the client will respond and say, "I don't want to pay less, what I want is to use my phone when I need it."

Figure 14: Maslow's pyramid.

Let us think about Maslow's pyramid (see figure 14). The pyramid is a theory of motivation that tries to explain what drives human behavior. The pyramid consists of five levels arranged hierarchically according to our desires: physiological, the need for security, social needs, self-esteem, and, at the last level, self-actualization needs. A product without utility cannot hide that its foundation is too weak, in the same manner in which someone who has nothing to eat cannot dream of self-realization.

I recently visited a company that needed to rethink its sales strategy. They had a great product from a utility point of view. When presented to potential clients, those who captured its power in total extent were amazed. They instantly screamed that they loved him. But this was the problem: only those experts who could understand the product in-depth chose to buy it. These were normally in the minority, had no decision-making capacity, or required another department's budget. The software was specialized, niche, designed for support areas. Only those who worked day by day on those specific tasks understood its functionality. Most times, the decision-makers were others. For some of them, presumably superfluous aspects such as allowing the graphic interface to be adapted to corporate colors were essential. "It has to be in turquoise blue, *absolutely.*" As a high command, your concern is with the brand. On the other hand, for creators who composed a leading product capable of solving complex technical problems, visual chromaticism could pass by. Such a mistake. Everyone perceives quality *according to their definition*.

When we talk about innovation it is worth bearing in mind the teachings of behavioral economics. This discipline has won two Nobel prizes in the first two decades of the 21st century, in 2017 with Richard Thaler and 2002 with Daniel Kahneman. The work in this field by the Canadian Hersh Shefrin is also very notable; and Robert Shiller's in behavioral finance on the market overreactions to the news.

Our judgments are governed by two systems of thought: one that is fast, intuitive, and based on emotions; and a slow reasoning-based system. We dream of living installed in the second of the systems. We are conscious, reflective, our premeditated actions ... The truth is that the first takes a more important role than it might seem. It originates feelings that are a source of beliefs. Beliefs are not always harmful, they allow us to make mechanical decisions, because we already know that. We don't need to recalculate everything on a day-to-day basis, which is practical. When we drive, we don't focus on every stomp to the clutch. But if they are mistaken beliefs, we find it difficult to correct them. That is why driving on the other side of the road in certain countries, or switching from automatic to a standard gear change is so laborious. As a result, a primary conclusion of behavioral economics is that we make choices based on approximations and conditioned by irrational aspects, those which govern the first system of thought.

Kahneman described these two systems extensively, although the phenomenon was partly known before his work.

Today a similar view associated with the evolution of humankind, called the triune brain, has become popular. This idea names each system of thought according to a physical section of the brain: reptilian, limbic, and neocortex. This was a beautiful analogy, surely more poetical than Kahneman's "one" and "two" systems. The problem is that some did not think it was a metaphor. Until today, it is often heard that we have a piece of inner brain that comes directly from lizards and that, from time to time, thinks for itself. Perhaps we should point to Carl Sagan and his 1977 book, *The Dragons of Eden*, where he talked about this. He is not to blame, however: Voltaire did the same with the famous story of Newton and the apple. We love stories, metaphors, similes. But the truth is that Paul MacLean, the original author of the triune model, always referred to it hypothetically and metaphorically, and the theory of the triune brain is not accepted in the scientific community.

Be as it may, if we do not stop and require the reflective system to work more often, we can make costly choices. We call these kinds of grim decisions cognitive biases. Let us take a look at a classic example that comes from the work of Kahneman and Amos Tversky himself towards the end of the 1970s:

> *Which of the following options would you prefer?*
> *Case 1: A sure win of $250, or a 25% chance of winning $1000 and a 75% chance of winning nothing?*
> *Case 2: A sure loss of $750, or a 75% chance of losing $1000 and a 25% chance of losing nothing?*

If we calculate the expected value—summation of the probabilities multiplied by the returns—both scenarios are the same. But Kahneman and Tversky shrewdly observed that human behavior differs in win-loss scenarios. Our risk aversion is greater in cases of loss. A majority would thus choose the first option in the first case, securing a profit, while taking the second option in the second case, risking a larger loss as long as they kept some hope of not losing anything. People who have invested in the stock market or traded intraday recognize this phenomenon: most stockholders are reluctant to cut a losing position and extend it until bankrupt. But as soon as those same stockholders see green numbers, they give up their patience to close the position so they can reap the benefits. This practice returns quick, small profits and long, deep losses. For not respecting basic principles of monetary management and expected value, between 80% and 95% of investors lose money trading, mainly for psychological reasons. This phenomenon happens because humans are less receptive to a certain failure scenario. We are more inclined to risk the possibility of falling a little more.

Similarly, people will tend to find the price of a product that has been downgraded from a full configuration for a discount more attractive than the opposite (i.e. starting from a basic product and charging higher) even though the final configuration and price are the same. This behavior, apparently irrational from the point of view of classical economic theory—according to which, the homo oeconomicus always makes rational decisions—is precisely how our mind works when it comes to buying.

What does this tell us? The Nespresso example teaches that we don't need something new to have a successful enterprise. The works of Kahneman, Tversky, and MacLean show that humans do not consistently make judicious choices in economic scenarios. Having an excellent product does not guarantee sales if we do not know what people want. Larry Keeley talks about at least ten different types of innovation:[1] in the profit model, in the network of contacts with which you can integrate your offer, in the organizational structure, in the processes, in the product's functionality (which may generate previously unknown needs), in the packaging (how you complement it with accessories), in client service, in the sales channel, in the brand, and in the customer engagement. We will comment on several of these methods of innovating throughout the book. There are oceans full of alternatives on how to innovate, even if your product is as simple as a blank sheet of paper. When you think about innovating, attack all five senses. Think about sight, taste, hearing, touch, and smell and then think about what you don't feel: the heart, the feelings, the social status. Design in your head a story for your consumers; imagine advertisements. When we consider ramifications, there is a universe to explore beyond functionality and the product itself.

OPEN INNOVATION

In the first section, we talked about the shared intellect and its benefits. One way to take advantage of it, as in the Linux and Wikipedia examples, is open innovation. A few years ago, Bill Gates referred to Microsoft as an "intelligence monopoly." Bill Joy, the co-founder of Sun Microsystems, sarcastically replied that "no matter who you are, the smartest people work for someone else." This story has been transferred to the collective imaginary as Joy's Law, although Gates was never involved and the story is apocryphal.[2] As a matter of fact, Gates himself is not exactly an admirer of intelligence tests.[3] Something is certain: it is impossible to monopolize talent or ideas—we cannot have them all. Traditionally, firms have managed innovation behind doors with research projects that came from the knowledge and instruments of the organization itself, believing that they also had a monopoly on intelligence. Open innovation just means cooperating with outside organizations or professionals. This is easy to say but not so much to execute it. Several cultural barriers arise, besides the habitual distrust of humans to share their grand ideas and knowledge. In this context, universities and research centers become relevant.

In the United States, we find the SBIR program (*Small Business Innovation Research*), which brings together funds from various agencies including agriculture, commerce, defense, education, energy, and health, among others. Firms like Qualcomm or Symantec have benefited from this program. In Europe, we have the Framework Program, baptized in its eighth edition as Horizon 2020, with a dedicated budget of around 77 billion euros. Horizon Europe, its ninth edition (2021-2027), maintains a budget of 94 billion euros. It concentrates all European innovation aid from large and small companies, which generate consortia and value chains with all kinds and sizes of organizations. Horizon 2020 is the largest research and innovation program in the EU. Between 2014 and 2016, 115,235 eligible proposals were filed for Horizon 2020, requesting a total financial contribution from the EU of 183 billion euros. This represents nearly 400,000 applications. In total, 13,903 grant agreements were signed, with an EU grant of 25 billion euros. Almost half of the participants are SMEs, which receive around 20% of the bag. Between 2014 and 2016, H2020 allocated almost 900 million euros to SMEs out of 2,319 concessions.[4]

It is not even necessary for companies to go in search for public funds. In Canada, they hold the Coopérathon (cooperathon.ca) every year. It is the largest open innovation event in the country, in which thousands of individuals take part. The event lasts for almost two months, from October to the end of November, and takes place in in five cities: Montreal, Quebec, Toronto, Waterloo, and Shawinigan. People use the Google Design Sprint methodology,[5] intended for rapid prototyping in one week, and participate in a crazy collaborative marathon. Coopérarthon has six tracks: finance, health, education, energy, the environment, and agriculture.

But why open innovation? "We believe that it is impossible for an organization to have all the best ideas," says General Electric Appliances CEO Chip Blankenship, "and we strive to collaborate with experts and entrepreneurs anywhere they share our passion for solving some of the most urgent problems in the world." GEA has its own growth strategy, (Ecomagination) and its online and physical community dedicated to the design and construction of household gadgets. In Spain, the optician Alain Afflelou promotes an active listening program for its clients through the online community ideas4afflelou.es (in Spanish only). Several business ideas have come out of this program—for example, interchangeable frame lenses, which offer the possibility to vary the style of glasses periodically, or a frame customization machine that gamifies its stores.

In Latin America, the Brazilian Natura started its open innovation program in 2006.[6] Ten years later, it had thirty deals with universities; more than half of its portfolio of cosmetic products came from these agreements. He structured a specific department, which was rather modest—about 250 employees in 2012. It launched the Natura Campus to consolidate this relationship with the academy and created other internal platforms that made it possible to sign alliances with non-academic professionals that foster this culture of innovation.

Searching for new ways to ally is also innovating.

THE ESSENTIAL ROLE OF THE GOVERNMENT

Open innovation is fueled by a close collaboration with the public sector.

Why has there been so much technological innovation in the United States in the past few decades? The answer is simple: The world's largest venture capital investor lives there. *It is the United States*. The assumption that the role of the public sector is limited to motivating innovation led by the private sector (through subsidies, tax reductions, pricing, technical standards, etc.) is as false as it is dangerous. This idea seems to ignore the plethora of examples where entrepreneurial effort comes from the government itself.

Take the example of the smartphone, the most important technological gadget from the beginning of this century. Compared to previous generations of phones, a smartphone is characterized by a series of additional functions that give it the ability to do many things besides executing phone calls. The enabling element of all these functions is the Internet, which smartphones capitalize like no other device. The development of the Internet began in the 1960s as part of a military project commissioned by the United States federal government. The aim was to build a new robust communication system with fault-tolerant architecture in computer networks. The pioneering network was called ARPANET, taking its name from the public agency that financed it—the Advanced Research Projects Agency, ARPA—along with the Department of Defense (DOD). The first message through ARPANET was sent in 1969. In parallel, initiatives were emerging in France and Great Britain, such as the NPL network. From a French project called CYCLADES and supported by the *Institut de Recherche en Informatique et en Automatique*, Vint Cerf and Robert Kahn took several concepts to propose: one was the current TCP/IP protocol, introduced in 1983. At the time, the project was also being funded by the National Science Foundation (NSF). With these inventions, computers could already connect and receive information—some through a Telnet client or protocols such as FTP. But as for the Internet as we know it today, we need to thank Tim

Berners-Lee. He gave us the web pages identified with URLs (like http://www.name.com), using the HTTP protocol and interconnected through hyperlinks—and the first Internet browser, too. Berners-Lee was then working at the CERN research center in Geneva with European public funds. It is 1990. Decades have passed since several countries in parallel began to work on the various technologies that make up the crossword that the internet was then. It is only at that moment that private initiative arrived en masse, taking less than ten years to produce the first economic bubble on March 11, 2000: the Dot-com crisis.

The public sector would give us even more: the first general-purpose explorer, Mosaic, developed with NCSA funding at the University of Illinois at Urbana–Champaign. The first private browser, Netscape, would not appear until 1994—Netscape came from the experience of its creator, Marc Andreessen, in the NCSA itself. There is always a link to the public sector. After gaining most of the market share in a few years, Mosaic was licensed to create Microsoft Explorer in 1995.

Countless services perch over the Internet. We already talked about PageRank and the Google search engine. It appears that Page and Brin's original work was also funded by a grant from the National Science Foundation.[7] Another significant technology is GPS that allows us to use such services as Waze or Google Maps. It was started by the Department of Defense in 1973 as a US public project and owned by the United States Air Force. Today, it is the American taxpayer who still pays for a service enjoyed worldwide. All program funding comes from general tax revenue[8]—which means about 1,500 million dollars annually.

"How could I search for a contact in my agenda? I know! Sliding with my finger." During the presentation of the iPhone in 2007, the attending public screamed in amazement at the demonstration of the new graphical interface by Steve Jobs's team. But touchscreens are old and of public origin: the first to describe a capacitive screen was Eric Johnson of the English research center Royal Radar Establishment in 1965. A few years later, CERN engineers Frank Beck and Bent Stumpe developed the first transparent touch screen. Touchscreens combine several technologies funded by the Department of Energy, the CIA, the NSF, and the Department of Defense.

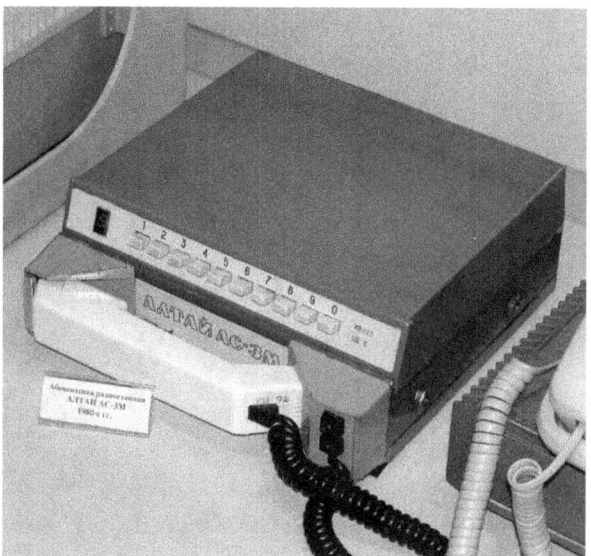
Figure 15: an Altai phone from the 80s.

The lithium battery comes from a project of the Department of Energy,[9] the microprocessor from DARPA.[10] The first generation 0G mobile communication system is from the Soviet Union, developed in the late 1950s and used since 1963 in several cities in the country, with a device called Altai (see figure 15). The first 1G network appears in Japan in 1979 thanks to the governmental company NTT, privatized six years later and today the fourth-largest telecommunications firm in the world. More than half of the ten largest telecommunications companies in the world in

2019 were or are state-owned: China Mobile, NTT, Deutsche Telekom, Telefónica, América Móvil, and China Telecom—the rest receive strong subsidies.

The phenomenon replicates in all industries. The pharmaceutical company is heavily subsidized by the NIH; the solar panels come from a project of the Department of Defense; the batteries that power Tesla's electric cars, too. Tesla has received about $4 billion from the United States government. Although it may seem like a lot, it is less than what non-electric car manufacturers receive.[11] Basic research involves decades of investment. Meanwhile, venture capital investors usually have no patience beyond five years. They typically wait until all the initial risk has been taken *by the government*. The most revolutionary technologies are also the riskiest. This does not exempt the private sector from merit and responsibility. Remember: basic research, development, and innovation are all necessary. Without public investment, there would be no basic research necessary to develop most of what is changing the world today. Without innovation, we would not have such well-designed products according to the needs of clients, who are, after all, who end up paying.

Let us remember. Human beings are fascinated by stories— it does not amuse us to see Newton locked up twenty years between formulas; we prefer an apple falling from a tree and causing a sudden quantum of genius. Let us look at the example of the largest integrated circuit production machinery corporation, the Dutch company ASML Holding NV, which has received public finances since its creation in 1984.[12] Today, it has an income of 11,000 million euros, over 23,000 employees, and dominates 80% of the market share. His story, told by the firm itself, meets all the literary requirements:

> *it was a matter of hard work, sweat and pure determination against almost insurmountable odds (...) It is a story of individuals who together*

achieved greatness

reads the ASML website. But later, they admit that the public funds collected after the 1986 crisis allowed them to be what they are now:

> *the competitors who had survived the crisis no longer had sufficient funds to develop.*

In the near-700 pages of Walter Isaacson's biography of Steve Jobs, the importance of public investment to make the iPhone available is never mentioned. We prefer the "Great Man Theory:" that historical advances are explained by the impact of highly influential heroes and individuals who, thanks to their charisma, intelligence, or political skills, had a decisive historical impact. It was popularized by Scottish author Thomas Carlyle. But as Herbert Spencer pointed out in his criticism a few years later, the so-called Great Men are, in reality, products of their societies, and their actions would not be possible without the social conditions that preceded them.

INTELLECTUAL PROPERTY

Innovation is a change that introduces novelties. It refers to modifying existing elements to improve or renew the whole. In general terms, innovation is achieving an end through knowledge, following a path that had not previously been followed. This means everything and nothing. What exactly do we consider a novelty? When it is something new for me? Does it need to be unique for all humanity? How much wisdom do you need to contribute? In innovation, context is important. Any application of knowledge that we make to achieve something new, and supposedly better, should be considered an innovation. But how radical the novelty is regarding the status quo, and the perimeter—new to us or the whole society—is also important.

The most radical innovations—the ones that differentiate us from all the others—are usually protected. To achieve this, there are some rules. Patents are granted on inventions in any sector, from washing machines to advanced nanotechnology chips—both physical products and processes. Sometimes a product involves several inventions. This is often the case in computers or advanced technology devices. It is worth drawing on the definitions necessary for an invention to be considered patentable.

The closest to an "international patent" is the Patent Cooperation Treaty (PCT). It allows protecting an invention in several countries at the same time. The PCT has five requirements: novelty, an inventive step, industrial applicability, sufficient description of the invention, and that it be protectable material.

The last two are formal aspects: the invention must be detailed enough to be reproduced by an expert, and the laws must allow it to be protected. Here are ethical-moral factors. It is not possible to patent certain therapeutic and diagnostic methods, but also harmless things, such as the source code of a software, which is not patentable in Europe if it is not in combination with a hardware complement.

Let us focus instead on the first three, which answer the question, *"What exactly do we mean by new?"*

1) **World novelty** means that the invention is not included in the current state-of-the-art on the entire planet. That is to say that the invention is unknown (not previously published or disclosed). To ensure this, the corresponding patent officer conducts a search during the patent acceptance process. If there are any already-existing public written documents of the application describing the invention, it will not be

considered novel. Talking about your invention on your blog can derail your chances of having it protected.

2) **Inventive step** avoids protecting inventions whose construction is obvious from what has been published, even if the final configuration has not been previously disclosed. The examiner comes to consider that something is evident by combining publications that describe each part of the set. In this way, we could not patent a "chair painted in turquoise blue," even if never described before, because it is made up of two known and simple parts: a chair and a turquoise blue paint. All inventions are infinite combinations of pre-existing things, however: the light bulb is a combination of inert argon enclosed in worked glass, where a tungsten filament is introduced and electricity injected to heat it. Everything was available, but nobody knew how to combine them before. It is common for many inventors to work in parallel on the same thing and end up in the same inventive step. We call this phenomenon "multiple discoveries."

3) **Industrial applicability** just means that the invention is useful. This is also subjective. *Industrial* here takes on a broad meaning, referring to any physical or technical activity. There are patents of the most varied and curious nature, however: a "greeting device" that automatically raises a hat while you walk which was patented by James Boyle in 1896, or "sunglasses for chickens," patented in 1903 by Andrew Jackson.

If we pay attention, much of what we usually call innovation is not about "patentable objects." Perhaps none of the innovations we plan to implement in our companies would be, unless we work in a research center or university. Vice versa, not-so-innovative things at first glance manage to be protected. Through utility models, we can protect the original uses of an existing object or configuration and fresh ways of manufacturing them. Utility models are also the only way to protect business models (or parts of it). For example, Amazon's one-click ordering method is proprietary. Aesthetic matters are also protectable with industrial design patents. The classic design of the Coca-Cola bottle is protected. The famous patent dispute between Samsung and Apple was based on the external and cosmetic design of the phones.

When we invent something, the most frequent step has been trying to patent it. A patent is a territorial exploitation right solely for its inventor, for a limited period of time — usually twenty years from filing the application. In the case of industrial designs and utility models, this figure varies depending on the country, normally ten years. Protecting is an intuitive idea in the current capitalist context, in which progress is always related to an economic benefit, although not so much in the historical context of humanity (as we saw in the first chapter). Almost all patent laws are scarcely two centuries old, even though the Venice Statute of 1474 is often cited as a pioneer. They are geographical rights: only valid in the country or region in which the application has been filed and the patent has been granted, in accordance with local regulations. Throughout different and strictly demarcated stages, it is possible to expand the places where I apply for my patent, increasing the payment of fees until I can no longer go back. In reality, a patent is a business case—a compromise between the high costs of their preparation and maintenance and the benefit that the exclusive exploitation in those territories will obtain.

A patent is also a strategic tool for businesses. There are patents called "offensive" and "defensive." Some companies register and save them, in case one day a competitor attempts to sue them. They will then search their portfolio for ways to fight back, a situation that usually ends in a license crossing.

This exchange of intellectual property between large firms is as tangled as the foreign debt between countries. Former Sun Microsystems President Jonathan Schwartz narrates on his blog[13] his fights with Bill Gates and Steve Jobs—the latter whom he calls a "patent troll." Patent trolls are unproductive companies that work to accumulate patents and earn royalties and lawsuits. In the title of his post, Schwartz paraphrases a famous quote attributed to Picasso, Faulkner, and TS Elliot: *"good artists copy, great artists steal."*

Are patents beneficial? The discussion about whether they are good or bad is old. The case of China is paradigmatic, with double-digit growth for decades without a respectful system for patents. Proponents often appeal to the moral need to reward the inventor for his effort. This stopped being true a long time ago. Today, patents must cite their inventors, but the owners are regularly the companies that bear the costs and those that profit from them. It seems clear that the importance of patents varies between industries: it is not the same with the pharmaceutical or chemical firms as agricultural producers. The debate about its social value has recently been revived, particularly with the awarding of the 2014 Nobel Prize in Economics to Jean Tirole, author of the book *Economy for the Common Good*. Tirole explained in his keynote address the problems of stacking royalties:

> *Biotech and software technologies are often covered by a multiplicity of patents of varying importance and owned by different owners. This 'patent thicket' is conducive to 'royalty stacking.' [...] it may be useful to return to medieval Europe, whose river transit was hampered by a multiplicity of tolls; for instance, there were 64 tolls on the Rhine river in the 14th century. Each toll collector set his toll to maximize his revenue, oblivious of what this meant not only for the users but also for other toll collectors. Europe had to wait until the Congress of Vienna in 1815 and subsequent legislations to see the removal of toll-stacking. High technologies are currently witnessing an evolution similar to river traffic*

> in the 19th century. New guidelines have been set to encourage comarketing of intellectual property through patent pools. Patent pools reduce the overall price of licensing complementary patents, benefiting both intellectual property owners and technology users.

Forbes magazine recently wondered if, beyond their benefit or harm, patents make sense: 97% of them never recover the capital invested and 50% do not complete their full cycle since their owners decline to continue paying.[14] It seems that the same digital forces that subjugate giant corporations to tiny startups are being applied to patents. A product stays on the market for an average of 18 months; is that worth protecting? Furthermore, the intellectual property of complex ideas requires a whole fortification covered by patents, since many parts must be protected. Although the article cites some use cases, the conclusion from Forbes's article seems clear: "Be original, keep innovating, and don't trust patents to protect you."

Patents are a business tool, not always meritocratic to inventors. Kirkpatrick MacMillan built the first bicycle in 1839. Gavin Dalzell owned the patent, however, and was credited with the invention for many years. Alexander Graham Bell obtained the patent for the telephone in 1876, but five years before Antonio Meucci had carried out a public demonstration of it, and ten years before Meucci, Johann Philipp Reis in Germany. Even earlier, Reis used the instructions that seven years previous, Charles Bourseul had published in an article in L'Illustration in Paris. The first patent for the phone was granted to Graham Bell after litigation, because that same day, February 14, 1876, Elisha Gray had filed another patent for the same invention.

Patents have also fueled one of the greatest myths related to innovation: the genius who changes the world with a brilliant epiphany protects it and becomes a millionaire. The famous rivalry between Thomas Edison and Nikola Tesla in some ways supports this: they both became millionaires, but Tesla was subsequently ruined.

In collaborative research, consortia usually protect their intellectual property. They put apart a budget for that reason. But companies that embrace open innovation are realistic about patents. If you manage to become the leader in a market that has great distribution and strong, long-lasting relationships with suppliers and retail channels, why would you care? Pursuing your copycats is costly, not to mention impossible, and they will always find a way to copy it by skirting the patent. There is only one thing that cannot be imitated: being the first to launch something. It's about selling, not protection. Being the first to market is a much better form of insurance than a patent. Patents are for everyone, but not everyone needs them.

CAPTURING INFORMATION: SOME TOOLS

Let us park the discussion on grassroots innovation and return to a closely related concept that we touched during the first pages: digital transformation. We have mentioned that digital transformation is a process of deep metamorphoses in an organization. It is leveraged in the technologies explained earlier, all supported by the transition from information to digital format. Digital transformation consists above all of breaking with the forms of work from the last century. Therefore, it is much more about strategy, making certain decisions, adopting a specific philosophy that permeates the corporate culture, more than something that technology can fix alone. It is useful, however, to briefly review some of the most common tools when digitizing an operation—some that any firm should be using. Let us remember that all modernization involves understanding what is happening around us, for which everything must be digitally registered.

If there are families of tools, it is because everyone, regardless of what we do, ends up performing similar tasks. It is essential to understand those common traces we share to be aware of where we are failing.

Let us start with the basics: we produce information. Whether in the form of books, magazines, letters, forms, manuscripts, personal notes, invoices, office records, photographs—everyone generates data and can take advantage of them. But just as we talk colloquially about good and bad cholesterol, there are good bits and the bad bits of data. Digitizing is not enough. If we do not have the information properly structured and stored, we can do little about it. Document management systems (DMS) are the first step to this problem: they help store, manage, and track documentation that has been transformed from a physical to an electronic format. They are the starting point for emerging digitization companies. At the moment, we are talking about documents: offers, contracts, videos or posters if we are an advertising company. As a result of implementing an SGD, it is possible to access any documentation instantly and keep track of what happens to it. Even if we combine it with a cloud system—as we will see later—it is easy to distribute content anywhere in the world. A well-known example of an SGD in the cloud is Google Docs: a "Microsoft Windows folder" that we can access from our browser. Google Docs allows us to perform various actions on documents online. Via email, it is feasible to distribute Word, PowerPoint, or Excel files. Dropbox is equally great for sharing and storing. Both are basic examples, because although they act as repositories of digital information or vaults, where it is possible to place our registers. The additional logic they implement is little. What differentiates a good SGD, one that is useful to us, is the range of actions offered on this vault: knowing who has accessed the document, when, generating permits, possessing traceability of edits, etc. Examples of more advanced tools include Custom Show, Clear Slide, or Zoho Docs. Other companies need tailored solutions, such as those that handle formatted documents—for example, engineering companies or architectural firms, which have documents that usually go through various approval channels, technical reviews, and that need to be signed at every step. Legal services departments, which handle sensitive documents in their chain of custody or require finding information in huge documents of hundreds of pages, also need a similar process. For them, Google Docs is not enough. These companies require solutions that usually incorporate additional functionalities that make them useful tools beyond acting as very expensive folders (barcode recognition, OCR scanning, automatic indexing, integration with active directories to regulate the access, etc.).

Regarding the security of access, a technology that is rapidly making its way is biometric patterns. It is essential to move from the world of physical documentations to the digitized versions of them. Although effective, these advances increase the fear of insecurity. We are used to signing paper contracts. Biometrics capture an individual's unique biological characteristics: iris and retina patterns, fingerprints, voice waves, the geometry of the earlobe, face, or hand. They collect not just our signature, but our gesture, too: some systems capture the firmness and the singular pressure patterns that we exert when signing—we start weak and end up pressing the paper more, for example. It is also feasible to analyze DNA, although, for obvious reasons, we will not see this in commercial contexts for now. More and more we will find more establishments where we will sign contracts with our fingerprint, where we will identify ourselves by looking at a camera. Likewise, it will soon not be necessary to answer personal questions during a call to customer service, because a system will recognize our voice while we are greeting the customer agent. When it comes to internal management systems, we will see more and more fingerprint and facial patterns to access documents.

Having documents like contracts or POs in digital format is an important first step, but is not enough. Two tools that are often confused help us go deeper in controlling our information: ERP and CRM. ERP stands for Enterprise Resource Planning. The idea is to have all the relevant operational information in a digital and structured format to improve the efficiency of business processes by exchanging it. We are no longer talking about documents; we are talking about data that can be combined, manipulated, or displayed. Data that express what we do and how we do it. When all this knowledge comes together, it provides a complete picture of what is happening within the organization. If there is a problem in one area, its effects become visible in other areas. This type of data highlighting encourages different departments to start working on the subject in question and take preventive measures—in turn, there is a greater rationalization of institutional processes. Organizations that do not have an ERP are often chaotic: they do not know how or find it difficult to calculate basic questions about their monetary operations. How much should the organization pay suppliers? How much should they receive from different departments? The calculations at the end of the accounting period are endless. Some modules are in charge, instead, of commercial aspects: the management of inventories, products, transport to warehouses, and purchases—and others in human resources, planning, and production. With all this, management receives reports every month, quarter or year, and makes decisions based off of these reports. Examples? The SAP suite, Oracle Net Suite, or Microsoft Dynamics.

In contrast, CRM software (customer relationship manager), offers general marketing, online campaign launch, and email distribution capabilities. There are modules to manage sales, orders, social networks, call centers, and complaints. The software helps in the management of potential clients, as well as in loyalty programs. Finally, the modules allow one you to look at the buying trends, so you can analyze them later. They enable you to gather, share, and categorize customer interactions even before they become customers, making it easy for business executives to project future sales. The logistics department verifies delivery addresses, while the billing department presents correct invoices. Salesforce is perhaps the most popular tool in this family.

ERPs are often confused with CRM systems. They are similar in the sense that they record and analyze information about what is happening in an organization. A mnemonic rule is to think of ERP as the orchestrator of data generated *inside* while CRM is its counterpart *outside*. Both are useful and relevant, but if you have to choose between the two, it helps a business owner to ask: is an increase in sales or greater efficiency required in operations? The answer should clarify the doubts. If the business owner wants to get rid of the skein of legacy or unintegrated applications, then ERP must step in, helping to order the processes. Alternatively, if the idea is to focus on marketing and attract customer traffic, CRM should help. Some solutions, such as Microsoft Dynamics, offer packages with both tools integrated.

These three systems help us put the house in place: having our documents digitized (DMS) and properly ordered; recording and possessing the ability to extract at all times the most relevant information from our operation (ERP), and managing our customer relationship (CMS).

Let us now see what happens in the outside.

The rise of electronic commerce in all industries is unstoppable. At the beginning of 2010, it represented 4.2% of total retail sales in the United States. In a decade, that number has tripled. In Spain, this indicator remains below 10% and in Latin America, it barely reaches 3%–4% in most countries, led by Brazil and Mexico.[15] The opportunity to gain market share is therefore colossal.

The reasons are transparent enough. More and more consumers are opting for digitized services worldwide; the penetration is increasing and the fear decreasing. Business to business (B2B) is also becoming more adapted to doing online.

There are e-commerce platforms for all tastes and needs. Shopify is suitable for both individuals and companies of almost all sizes, providing the company with an advantage. Do you know anything about programming web pages or an online store? It doesn't matter, Shopify requires no coding knowledge. The platform makes professional tools available to everyone: payment gateways, search engine optimization, and an endless cast of applications for every imaginable functionality. How did they do it? You guessed it: they opened their platform to third-party developments. They are now also a platform connecting online store owners and software developers specialized in e-commerce.

Shopify falls, for now, a bit short in capacity for larger companies, with potentially thousands of daily transactions. But BigCommerce goes hand in hand with Shopify in several ways: it is simple, convenient for all stores of any size, with glorious buyer-centric and customizable themes and templates, many of which are also developed by third parties. Despite this, it possesses advanced and robust features that makes it a leader among the largest corporations.

Magento is likewise a suitable option for large firms, but its use is more technical. Programming knowledge is essential here to properly manipulate the platform and adapt it to what we need, although it also offers templates, widgets, and other plugins. Sizable companies with sufficient funds should not find it difficult to hire Magento developers to create fully customized online stores.

Let us talk about some trends in the world of e-commerce to keep in mind. The first trend is the unstoppable rise of mobile phones as the primary tool for accessing the internet, above personal computers. In most countries, the tilt of the balance in favor of mobile use for computers occurred sometime between 2015 and 2016. In some, the tilt was even earlier, such as India in mid-2012.[16] This change forced online store developers to adopt a mobile-first philosophy: think about mobile users above all. The next step will be to purchase by voice, relying on virtual assistants from operators or phone manufacturers (such as Apple's Siri) or assistants the eCommerce stores will gradually develop themselves. Artificial intelligence and augmented reality tools will gain importance. The landscape of online sales and purchases will radically change soon.

On the other hand, the influence of social networks as advertising channels for products or services will increase. Facebook, Instagram, Pinterest, and contemporary entrants have already become visible platforms for various brands and will continue to gain further visibility. These websites are the new television: internet users consume two and a half hours a day, according to a survey conducted in 31 countries.[17] Those who were under 24 years old in 2019, the next consumer generation, spent over 3 hours on the internet.

This brings us to the last tool we will inspect: marketing automation. The online marketplace is never static, so an organization has to keep abreast of current volatile trends if it is to survive in the competitive arena of global trade. There is no point in taking over all aspects of a business manually if we want to invest time and effort in adapting our online store. The online digital marketing is still an unexplored and strange planet for anyone who studied marketing more than a decade ago.

Several pieces of software can help with these repetitive tasks. For publishing relevant content on social networks: MeetEdgar or HootSuite; for sending automated and personalized emails to clients: MailChimp; for kick-starting advertising campaigns: HubSpot—which is a true CRM adapted to the online world. These tools are highly effective and fast. If a business uses them wisely, they are powerful enough to turn potential purchasers into customers with no or little human intervention, thus saving time.

People who make purchases offline face all kinds of experiences. Some are pleasant and some are downright horrible. Depending on the experience, a shopper can continue with the seller or avoid the store in question forever. The opportunity is as great as the risk of not doing things well.

STUMBLE A HUNDRED TIMES ON THE SAME STONE

In July 2000, Florentino Pérez won Real Madrid's presidential election for the first time. His star electoral promise was to sign Luis Figo, a Portuguese man who played for his eternal rival, FC Barcelona. Said and done. That summer, Real Madrid became the club to pay the highest price tag ever for a soccer player, signing a check for 60 million euros. A year later, they broke the bar again, paying 74 for Zinedine Zidane. The following year it would be Ronaldo Nazário, and in 2003, David Beckham. There was speculation about the new president's heavy investments in stars. Florentino became notorious for a speech in which he compared soccer players with productive machines (he was likewise the CEO of the construction company ACS). "When I buy a machine, I calculate what is the return that it will give me. Same with my players. They cost me so much, they return so much. We recover it by selling jerseys." We all laughed in response. Barcelona's president Rosell declared that it was impossible to redeem the investment with merchandising. Whether it was possible or not, the truth is that for the first time the acquisition of soccer idols was justified with a financial business case. For Real Madrid, or at least for its chief, the matter was simple: I will invest that amount of money and receive this in exchange.

Unfortunately, things are not that straightforward in the world of innovation. Let us take the specific case of the adoption of artificial intelligence, which we will talk about in detail later. A study[18] from the MIT Sloan Management Review and the Boston Consulting Group was performed on 2,500 executives, mainly American, published in October 2019. 40% of the companies reported that significant investments in artificial intelligence had not yet seen returns.

There is a noticeable difference between winning and losing adoption strategies. As you may have imagined, that difference has to do with everything except technology. To be fair, there is a technological component, but it combines with a strategic and a cultural one. Some companies have a vision focused on technology: they perceive artificial intelligence as a lifesaver that will come to solve their battered sales funnel, without stopping previously to analyze said funnel. There are also those instances where the CIO makes certain purchasing decisions to "digitize for digitizing" without taking into account commercial needs or a business case made by the departments, be they Product, Operation, or Sales. Finally, there are those who expect to see an impressive return with absolute certainty, which is usually the most common. They all tend to fail.

It does not matter whether we refer to artificial intelligence—a tool to improve SEO and digital marketing campaigns—or to the establishment of a research group for market perception. Although it may not seem like it, the foundations of success are the same. KPMG conducted a study of 400 executives and found similar situations. It is best for companies to start small and make sure there will be no resistance from the first stage. Big bang changes often encounter barriers. Most of the ideas proposed in companies go through the Overton window filter. The Overton window is a political theory that limits to a narrow range the views that the receiving public—in this case, a Steering Committee—finds acceptable. Before venturing to bring up any proposal, the vision of the organization should be clear. Iterating aimlessly is like a ship without a rudder.

In large enough organizations, if the CEO is not the main sponsor, it is best to leave the transformation for another day. In small ones, there must be a plan. The mission and aim of a company do not change in the long term, at least they should not. The strategy is the set of choices made according to changes in context and environment. The plans and tactics are instead flexible, iterative, and adaptive. Is there a vision-based strategy and a series of ploys that unravel that vision and filter it across areas?

Best practices advise starting with the client's needs, using data, focusing on doing less but better, applying agile concepts to iterate from the basics, thinking about simplicity, and partnering to share knowledge. Do we know our clients? Do we talk to them and do we understand what they need? Only after having the clear answers to these questions can we delve into the technological storm.

Far from majesties, the reasons why digital transformations fail in traditional industries are often rather vulgar. The human factor, the executive's agendas, fear, laziness, and other capital sins are usually much more frequent than sophisticated miscalculations when choosing and implementing cutting-edge technological solutions. Many young graduates feel amazed at the lack of common sense in the way organizations operate until they too adapt. Others quickly become disillusioned, finding that the situation is nothing like what they had imagined from world-renowned companies, and are increasingly seeking shelter in small but dynamic and challenging projects, trying different things. As a result, previous generations label these workers as whimsical, difficult to manage, narcissistic, and an endless string of pejorative adjectives. Perhaps part of it is true—that the last few generations have been the victims of poor education, the influence of ubiquitous technology, and immediate gratification that their parents don't understand. Perhaps it is also time for preceding generations to look in the mirror.

We suffer an explosion of bad leadership in the world: not incapable leaders, but misunderstood leadership. Understanding and accepting the concept and vision for transformation is not the same as committing to doing what it takes to succeed. Business leaders can have a great meeting and talk about the need for change and they can read a thousand books better than this one, hire dozens of agencies and consultants—but if they do not commit to a deep transformation, which necessary to achieve the vision and the strategic intention, the initiative will fail. And it fails 70% of the time.[19] We should understand that the education we apply in the workplace does not work. The focus leans increasingly toward the short term. The hierarchy is excessive; we lack understanding of what is coming. In the United States, 58% of managers have not received managerial training.[20] In other countries, the figure is higher: 89% of them believe that people leave the company for more money.[21] In reality, only 12% do. Meanwhile, 53% are unhappy at their job.[22] A sad truth?

START AN INNOVATION FACTORY

6

"If you want to make an apple pie from scratch, you must first create the universe."
—Carl Sagan, 1934–1996

The day Scuderia Ferrari named Joan Villadelprat chief mechanic of the team, he stumbled upon a dire picture. The Italians applied the same bohemian and artistic philosophy to the construction of their Formula 1 as to their lives. It was 1987, the eighth year in a row of Ferrari not winning the drivers' world championship. And despite Villadelprat, it would take another thirteen to do it again.

Notwithstanding his youth—he was in his early thirties at the time—the Barcelona-born had spent six seasons in the highest motorsport competition. He had worked his way through a couple of British teams, Tyrell and McLaren, who taught him something fundamental: to systematize the construction of his automobiles. At Ferrari, Formula 1 cars were mounted and dismounted on the spot, at every Grand Prix and practice session. All different. Villadelprat's first mission was to get his mechanics to manufacture two identical cars each time, an incredible achievement in Maranello. He applied his disciplined British experience until he obtained a repeatable method.

In the season before Villadelprat's arrival, Michele Alboreto finished the championship in ninth place. His best Grand Prix result had been fourth, not managing to finish nine times. In the following year, Gerhard Berger, their best driver, completed the championship in the fifth position, but also with nine dropouts. Then he finished third, only behind the unattainable Senna and Prost, with a Grand Prix victory, the first in years, and only five abandonments. Then Nigel Mansell finished fourth with two wins.

Despite not succeeding with Ferrari, Villadelprat was able to accumulate five drivers' and three manufacturers' world championships throughout his career. In his words, two of them were obtained by "the team of a sweater company with a ridiculous budget." He was referring to Benetton. Villadelprat was part of the team who won the drivers' world championship with Michael Schumacher.

Innovating with a method makes sense. Throughout the next pages, we will study ways to put order in our approach to "building Formula 1." Alice asked the Cheshire Cat which route she should take. He replied, "That depends on where you want to go." Since Alice didn't care much, the cat confirmed: "Then it really doesn't matter, does it?" Our types of innovation must be accompanied by their respective purposes and aligned with the company's overall vision and strategy. If we are not capable of this, we have a problem. We should also link them to the operational objectives of each group or business unit. Four tools will be very useful here:

1) **An unequivocal idea classification method,** i.e., a way of classifying them that everyone can apply without much difficulty.

2) A **blueprint or process map** that allows discovering weak spots in our contacts with the client, generally at the customer journey level.[1] Analyzing the blueprint, we can obtain an analysis of the **pain points** of the company to link them to different initiatives.
3) An organized system for **assigning objectives**, from business to individual, through those of each department. Later we will talk about OKR to discuss a specific proposal in this regard; OKRs are an especially convenient goal management method because it forces transparency of these goals at all levels. Beyond defining them, organizations must work on sharing and communicating them.
4) Some **filtering rules** would also come handy: what will we do, what we will not do—which is most important—and what can we consider and discuss internally.

Doing all of this is not an easy task, but it is the beginning of a well-formed innovation system and culture aligned with what we want as an organization.

CLASSIFY IDEAS

How do people devise new ideas? Many authors have set out in the quest of a common method, in the same fashion that philosophers pursued a universal ethic. There are probably stable factors when it comes to creativity. After each new scientific publication and another new product introduced on the market, both colleagues and competitors exclaim: "How stupid was I to not think of this before?"

Why *didn't* they think about it?

It takes people capable of connecting two points that did not seem a priori connected. Making that connection requires not only intelligence and training in a particular field, but also an act of certain courage and audacity. I would propose adding one more element: a safety mattress against the consequences. For a career in the corporate world, compliance works much better than creativity. Speculating out loud is dangerous, and every word must have been previously measured. Ideas exist, but they remain silent in the heads of their owners. The ones that do come to light are usually obvious connections between points, which many reach simultaneously. They are corollaries, not new solutions. They are just on the shore of the ocean of creativity. Individuals willing to go against the status quo are often nonconformists—hard to fit. Since his head is unconventional, his habits probably aren't either.

To innovate in the private business world, to embrace a culture of error is capital. A mistake should be allowed, a general sense of permissiveness towards stupidity reign. The wonderful thing about stupidity is that, in some cases, it ends up becoming reasonable: first, they ignore it, then they laugh, then they attack it, finally everyone embraces it.

Creativity is embarrassing. No one has a perfect idea the first time, and even if it is close, it needs to be polished. For every good thought, there were a thousand foolish ones that nobody wanted to tell in public. Anglo-Saxon cultures are much more likely to accept errors and foster a culture of experimentation. It is a pending subject of the Spanish-speaking world that is feasible to achieve nevertheless. Small trusted groups are necessary to obtain the best results. You must be able to say what is needed without overthinking it much. The group must have the confidence not to call each other by their last name. This process is exactly the opposite of what occurs in most works meetings. But implementing an innovation system will need all the attention and support from the general managers possible.

Ideas will come if we achieve this culture of error—then we can begin to classify them. In order to obtain a healthy innovation portfolio, it is best to begin different types of projects at the same time. A giant portfolio is not required, but keep in mind that we will undertake projects with a closer impact over time and others that will take longer to bear fruit. Both methods are necessary for success. Let us recall the basic research, development, and innovation classification - similarly, we can have projects fit in any of those levels.

There is no shortage of adjectives for innovation. For the shorter term: sustainable, incremental, architectural (a term used for market diversification), continuous improvement programs, organic growth initiatives. Or disruptive innovations, new growth initiatives, blue oceans, for the longer term. But strategically speaking, all innovations fall into one of two types.

In the first are those that expand the current business, either by enhancing existing offers or products (usually in terms of functionality or performance) or by improving internal operations. These innovations are sometimes triggered by comments from current customers. There are issues with this concept, encapsulated by Henry Ford's infamous quote: "If I had asked people what they wanted, they would have said faster horses."

The second are those that generate new growth by reaching unexplored customer segments or markets, often through novel business models, or even creating needs that the consumer did not yet have. Regarding continuous improvement, radical innovation is usually accompanied by greater defects, incomplete products in beta phase or prototypes, with insufficient initial performance. Typically, they can only be targeted to a narrow niche market or group of pioneers. For example, the early mobile phones with integrated digital cameras were not as popular at the outset. The quality was terrible and no one used them; they were making their way, however, until completely displacing the digital cameras that dominated during the first decade of this century.

Our portfolio must encompass both types of innovation. This does not mean that everybody involved should work on projects of both kinds, but it does require that everyone know the difference. We should understand not only the categories that we define but also their objectives. Otherwise, new growth initiatives will demand results too early. If we dedicate ourselves to basic research, this classification of two subjects is insufficient. For example, European programs use a nine-levels TRL scale (technology readiness level). From the outlook of an organization that does not have a basic research department, this is excessive. For the simplified presentation that we are going to take, and for the needs of most organizations, two to four are sufficient.

Failing to design a diversified portfolio of innovation is often the result of widespread confusion between these two types. Let us call them from now on "radical innovation" and "continuous improvement." Large—and usually older—companies that have dominated a market for many years tend to lean toward the second innovation type while crashing down on the first. They say they innovate, and it is true, but always over the same thing. For this reason, they are in a steady risk of disappearance. In this situation, radical innovation offers a better life for these companies. It is a process that enables a change of strategy in the future, creates lifeboats, and allows the generation of unexpected businesses.

Two great families. Not so many. Several organizations with a short-term vision exclusively follow a continuous improvement program. Others panic: "We want to innovate because we are losing money and market share by leaps and bounds." They need something radical and they need it now. When they get to that point, it is probably too late. So, having a level portfolio is essential from the beginning.

There is no single approach to categorizing projects. We simply need a way to differentiate innovations in our present activity versus innovations that will bring new markets or customer segments. A fairly popular model is the three growth horizons model, proposed by authors Mehrdad Baghai, Steve Coley, and David White in their 1999 book *The Alchemy of Growth*. According to this method, projects are classified into three levels: Horizon One (H1), Horizon Two (H2), and Horizon Three (H3). At the first level, the focus is on defending and expanding the company's core business. This includes all the incremental improvements of systems, processes, logistics, customer service, and so on. The H2 horizon brings the creation of emerging business opportunities that allow growth in the mid-term. These opportunities are businesses whose degree of maturity allows us to glimpse a not-too-distant implementation. In H3 we grow seeds, the crazy ideas that we only want to pilot, the options that may be tested in the medium term, and allow growth in the long term. A rule of thumb for our portfolio is to have the following distribution: 70% H1, 20% H2, and 10% H3, but it must be adapted to each need.

Does this contradict what was previously stated, that we only have two levels of innovation? No, it does not. It is not necessary to adhere to this scheme. Procter & Gamble[2] uses a four-types model. Citibank[3] uses three, similar to the horizons, but with a different name. It is perfectly possible to work with only two types. Basically, the three-horizon model subdivides the second type, projects that will bring us future growth, into two: one more feasible and the other more crazy and futuristic. No matter what strategy we use, we must be clear about our taxonomy and at least one class of projects that deals with forthcoming innovations. We may find ourselves with doubts when classifying. So, an explicit definition is recommended. Still, we will encounter dilemmas: can this project be classified as H2? Or is it rather H3?

For simplicity, we will follow the process with a system that has only two classes: we will call them "continuous improvement" and "radical innovation." When you are starting, it is recommended not to complicate the process much more.

Once we have our classification method defined, the next step is to devise a path for our ideas: who will execute them? What budget will they be assigned? Properly categorizing a proposal and linking it to a certain impact and objectives is interesting. Now it's time to work it. But first, we must make sure that our ideas, especially those of continuous improvement, really attack the problems of the organization. How do we do this? We take a look at the process maps and pain points.

MAPPING PROCESSES AND PAIN POINTS

In November 2015, McKinsey published an interesting article[4] on the impact of digitization. It presented some impressive cases of improvement obtained after digitalizing certain customer journeys. What is a customer journey?

Let us think of any organization, in this case a travel agency. What characterizes and typifies this firm as a "travel agency" is its expert knowledge. They understand what to do to manage, organize, plan, and execute travel-related purchases for a client, brokering with airlines or car rental agencies. They know whom to contact in each case, say groups that set up excursions to natural parks. Just the kind of entities that are unfamiliar to a construction company or a fumigation chemical factory.

On the other hand, our agency houses employees, computer systems, processes, and business culture, like most of the others. It is externally related to suppliers, customers, and the government regulation of its territory. Their business model may not be very distant from that of other groups with a radically divergent activity: it collects specific elements, in its case a train ticket and a hotel reservation, and adds a margin on the price they get, perhaps downgraded as an intermediary. Knowledge differs, but what they do, not so much.

Under a certain prism, all organizations are alike.

For this reason, there are operating models, also known as process frameworks, or blueprints. Although activities may seem very different in the far distance, companies have similar internal behavior in many cases. Most of the corporations initiate their interaction with the client through advertising campaigns: concept, campaign, and lead generation. Not all do it the same way. In business schools, it is common to review the Inditex case study. This Galician company managed to achieve world renown without investing in advertising. Well, one part is true: Zara never had a regular promotional strategy. They did not rent display space on airport panels or billboards, they did not have radio or television spots, nor distribute leaflets. What they have done instead was worked with Instagram influencers and bought pages in newspapers to advertise their sales. Inditex does marketing: its main asset is right-located stores, excellent interior retail design, and they were pioneers in storefront layouts. All societies need, at least, to think about how they are going to reach their customers, even if they run other methods.

Similarly, most businesses receive a purchasing order (PO), be it a formal document or in the form of someone at the cashier waiting to pay for a pair of pants, or even a request via mobile phone, as when we hire Rappi or Uber. Those POs result in payments. A reiterated service generates an automatic payment, such as a monthly subscription. Almost all companies require Customer Service: We receive questions, requests for changes, complaints from our clients. Then we try to offer solutions. Sometimes Customer Service is a complete department—other times the sales team does some after-sales work. Occasionally the same owner or founder is responsible for responding one by one to all complaints, especially if she runs an SME. Virtually all firms must generate retention or loyalty strategies or, eventually, have workflows that end the relationship. Some sectors, such as telephone enterprises, are recognized for the difficulty clients have in terminating their contracts.

An *operating model* is just an abstract and visual representation of how an organization delivers value to its customers. It is a drawing of what they do and how they do it. It is very important to produce in writing and keep these maps in order to generate an effective strategy. Without being clear about how a company operates, digitization strategies have a high risk of failure. An operating model breaks the complex organizational system down into small pieces and shows how they work. This model will help different participants grasp the whole picture, identify problems that cause poor performance, and understand what changes are needed.

There are pre-made models, such as the Service Operating Model[5] for services or the eTOM[6] in the telecommunications industry.

Our drawing may be general, limited to large macro processes, or more specific, to describe in detail each of the interactions that our clients have with us in each of the places or systems where they occur. These places are called *touchpoints*. The customer journey is the whole course that the client traces from the moment he decides to act—even before she chooses to—until the end.

The summation of these journeys is what defines the consumer experience and what, in many cases, differentiates successful brands from those that are not. This is why these maps are so important. Today, travel agencies have disappeared in a high percentage, and those that continue to exist have been sheltered in niche markets: luxury, private experiences, or large corporations with huge volumes of corporate travel. A majority of clients have discovered that they have other channels of access to the same services in a very simple way: search engines for better offers, reservations directly online, etc. The same companies also charge a commission, but I can consult them at any time. There, customers are free to explore, compare, and discover alternatives. Browsers can find recommendations for themselves and execute purchases with their card and without embarrassing themselves by asking too many questions only to decline purchasing a product. The customer journeys of those looking for a trip have changed. Traditional tour agencies have been either digitized or neglected by the clients.

Why is it so important to keep our processes well mapped? Because it allows us to define the criteria to decide what things we want to improve, digitize, or leave as they are. These decisions are better off when we have a complete vision of the operation. We can quantify, "How many communications with the buyer are made due to a change request? What are the preferred channels of our customers to make a return? Are they mostly carried out in physical stores? It seems that our client is looking for personalized treatment." It is possible to characterize our processes through several variables: volume of interactions, associated costs, related income, and many more.

In addition to describing the operational and support processes, it is requisite to know who is responsible for each of them. These owners are essential to ensure that the documentation reflects reality and represents improvement and value. It is fundamental that each process has an owner and that takes care of it. She is responsible for the definition that we have for the entire company. Once we have a set of customer journeys and processes consistent with our strategy and with the planned investment and return, we can start reimagining them, improving them, and then, only then, digitizing them.

I hope this helps to gradually highlight why technology is the last step. If we install tools over blurred operational maps, it will be a failure. If we rely on partial views or opinions—and we all have them—the process will fail, too.

Each of these journeys is made up of certain steps. Some of them are prone to be improved with technology. In the purchasing process, the client may be legally required to sign a document. We could propose a digital signature. This would improve the buyer experience, speed up waiting time, and avoid paper and custody expenses. You can save on transportation costs and enable online shopping without leaving home. If it seems reasonable, we can make a business case and involve the client to understand if that solution will be accepted. We should experiment. Then implement the best solution. Eventually, the process automation could further develop by introducing biometric patterns, so that we reduce fraud and obtain additional intelligence that is then reused to personalize advertising campaigns or have more accurate information about our customer base. It is like a great snowball that has been falling, in which technology enables more and more functionality in every new step. But it all starts from mundane, non-technological questions: where do we want to start, what hurts the consumer, what troubles us interacting with him, what is our long-term vision?

The customer journeys are translated into a number of internal processes, invisible to the buyer. When a client walks the "buy basketball shoes at the store" path, that can include the internal process "registering the sale of the item from inventory, so that it can be replaced later."

An internal process gathers tasks with the same purpose and defines how to do them. Therefore, it clarifies who is responsible for the actions of the process, where they want to go, and how each of those tasks or steps help to achieve that aim. What is the distinction between a process and a project? One imagines its complexity. A project is a great civil work, while a process perhaps means buying office supplies. In truth, the primary difference between a process and a project is that the latter has a clearly defined beginning and end, while processes are usually continuous tasks, spread out over time. But some processes can become as complex as what our imagination draws as a project. Digital transformation is concerned with making certain processes more efficient, fast, or enjoyable through digitization. If our current processes are not well explained, it is silly to aspire to succeed in either case. Innovation in itself is closer to being a process than a brilliant idea born of a privileged and bohemian mind.

I think there are some fundamental questions we should ask ourselves when thinking about defining a blueprint or a process improvement plan:

1) How are we going to register the way we work? Will we use a special notation or language, just text on paper? Shall we use drawings, illustrations, or diagrams? Will we hire an expert? Will we acquire new software? The goal would be for any newly recruited individual to know what to do in each case, simply by consulting the documented guidelines.
2) How are we going to make sure that everything runs correctly and in the proper order? Will we have a quality system or department? Will we implement a tool in this regard, such as a BPM?
3) Once the plan is complete, how do we improve it? That is, how we implement a continuous improvement system for our internal processes?

A BPM (Business Process Manager), in case you were left in doubt, is a tool that forces us to follow the best practices in the execution of processes. It allows, for example, to launch flows by mail, so that those involved are automatically notified of what is happening with their tasks. In many cases, it is a good idea to implement a BPM, since it is the simplest tool to digitize communications.

ASSIGN GOALS

The third element is an organized system to designate objectives.

An original way to design and communicate the objectives of an organization is OKR[5]. It is a collaborative protocol invented at Intel but popularized since its implementation at Google at the beginning of the century. Let us see how it works.

An *objective* (O) is simply what is to be achieved. We write it in a simple but well-defined phrase. Goals should be understandable, concrete, action-oriented, and ideally inspiring.

The *key results* (KR) are indicators that monitor and quantify numerically if we are reaching that objective. They should be specific, of a determined duration, aggressive but realistic and, above all, measurable and verifiable. A good KR has a well-defined magnitude and unity. In other words, there are no gray areas in successful key results. That means no "grow a lot" or "try to lose just a little." KR come with figures

---

[5] In addition to John Doerr's book, *Measure What Matters*, a unanimous reference to enter the world of OKR, I can recommend the work of Felipe Castro, a specialized consultant based in Miami and who has countless free resources on his website http://www.felipecastro.com

and measurements, and generally more than one for each objective.

Let us see a simple example:

OBJECTIVE	o Delight our customers.
KEY RESULTS	o NPS of 52 or higher. o Google Play score of 4-star or higher

The objectives are the beautiful, inspirational sentences, the horizons you want to reach—what all companies communicate and publicize in their internal quarterly meetings. The key results are earthy, boring, and metric-based, but also what differentiates honest organizations from those that are not.

The definition of objectives is nothing new and reminds us of the SMART wording of the 80s—specific, measurable, assignable, realistic, and temporary. We can identify them in almost any business. With the key results (KR) we already become more visible; KR make it difficult for us to cover our sorrows. It is less frequent to find them. Okay, but this is nothing from another planet. What makes the difference is how this large-scale goal expands to all members of an organization. Employees have well-defined actions ("deliver their order to the distributors every day at 9:30"), transparent to the entire organization, and easily linked to the vision. Those same actions are reconsidered in short periods.

This is the hard but really valuable thing about OKR: transparency, dynamism, alignment, and adaptation at every level. OKR take from the Agile philosophy that we will explain in a while. When I speak of transparency, I mean that the objectives and their fulfillment are completely public, from the CEO to the last apprentice. When I speak of dynamism, I

184 | Start an innovation factory

speak of weekly, biweekly, or quarterly reviews—not yearly. When I speak of aligned and adapted, I mean that anyone should be able to understand, simply by reading them, how their OKRs relate to those of their supervisor, and these successively to theirs, up to the highest level.

OKR ensure that everyone in a firm focuses efforts on the same essential issues. Successful organizations focus on initiatives that can make a difference now while deferring less urgent ones. A short three-month horizon slows procrastination, increases focus, and pushes us to do more. And not only this: regular checks (preferably weekly) are also essential for the same reasons. OKR aligns and accelerates the organization without the need to rework ideas, spend more and more hours, and lower the morale of the staff. Management must commit to those choices with words and deeds. The objectives are transparent, they cannot hide behind speeches that will have been forgotten in a few days. By staying there, steadfast behind the highest-level OKR they have defined and published themselves, they are giving their teams a compass and a starting point. Transparent and chained objectives also expose redundant efforts, saving time and money, and becoming by themselves a form of communication. Criticism and corrections are in public view. Transparency generates collaboration: when people see how their goals are connected to their colleagues, they can contribute more significantly to the company's success and see the general consequences of their actions. It is feasible to see when someone needs help and offer support. Public goals are more likely to be achieved than goals kept private.

Keeping a tree structure for our goals increases motivation and facilitates cultural assimilation. Newcomers adapt earlier. An effective OKR system links individual goals to the group's broader mission—its purpose. This clarity translates into job satisfaction for the entire organization. Everyone can identify not only their duties but the impact they have on the

What we talk about when we talk about innovation | 185

company's whole. In an apocryphal but beautiful story, we find President Kennedy visiting NASA in 1962. He approaches a janitor and asks him what he does. "I am helping humankind get to the moon, sir." Sure enough, he was doing it. If you ask your team, "What are the organization's top priorities?" they may be aware of them, but find it difficult to extract from their specific daily tasks unless they are provided with a map. OKR are that map. And it doesn't just work for large organizations. In smaller startups, OKR help founders drop tasks that are essential only in the beginning: accounting, payroll, administration, etc. They assist them in focusing on the product, the strategy, the team—the high-ranking goals. In other words, they keep them away from micromanaging and multitasking. On the other hand, in large companies, it is common to find several people working on the same task without realizing it—or what is worse, being perfectly conscious of it. Lack of alignment is the number one obstacle between strategy and execution. Healthy organizations encourage some goals to come from the bottom up, while others are defined by leaders. In other words, a healthy OKR environment strikes a balance between alignment and autonomy, shared purpose, and creative freedom.

Not all objectives are the same, nor do they have the same horizon, in the same sense that not all innovation projects have it, as we saw when talking about the three-horizon model. Google, for example, divides its OKRs into two groups, like other organizations divide their innovation portfolio into three or four types. On the one hand, there are the committed objectives, the equivalents to H1, which are immediate and achievable, thus "continuous improvement": product launches, near-in-time purchases or alliances with partners, etc. These must be fully achieved. Then there are the aspirational goals, those that correspond to radical innovation, equivalent to H2 or H3—broader, riskier, and more forward-thinking ideas. They originate at any level and aim to mobilize

Start an innovation factory

the entire organization. By definition, they are difficult to achieve. They are, in fact, met at an average rate of 60 percent. Google is clear that a success rate around that figure is satisfactory. This acceptance is interesting because it shows that Google is much more down to earth than other less successful companies with implausible strategic plans. Although we know that they will not be wholly completed, having ambitious objectives drives companies to ask key questions: what kind of institution do we need to be next year? Agile and bold, to break a new market, or more conservative and operational, to reaffirm our current position? It forces us to think beyond.

Similar to the use of technology in digital transformation, the implementation of OKR is relatively the easy part. The piece of software or the methodology we are implementing does not matter as much as the human touch and the willingness to carry on.

Large institutions often have an objective management system already, which is typically managed by the Human Resources department. The annual objectives are loaded there, established, and then forgotten until the time of review and evaluation arrives. OKR instead are throbbing data-driven beings. They should be tracked, revised, or adapted depending on the circumstances. In huge companies, the implementation of specific software becomes necessary. More and more organizations are adopting dedicated, cloud-based OKR management software such as Atiim, Wrike, or 15Five, where users navigate a digital dashboard to create, track, edit, and rate their OKR, as well as view their connections with the OKR of others. Websites like okrsoftware.com also offer comparisons. But remember that at the end of the day, what matters is wishing to do it. Some companies choose to adapt their existing software or download templates to use with Docs or Excel. Only a transparent, collaborative, aligned and connected organization achieves this. If you get transparency through an Excel sheet because you still don't have a budget for software purchase, that's secondary; you will end up getting it.

In recent years, it has become fashionable to call Human Resources departments "People," under the assumption that we are not just resources to use and throw away. The message seems to imply that the new relationship between bosses and their employees will be, after the label transition, more humane. But what the trick reveals is just the opposite. Nevertheless, whether we like it or not, individuals cannot be reduced to numbers. If a conversation is limited to knowing if you have achieved the objective, the context is lost. Continuous performance management is needed to explore some pertinent questions: was it an achievable goal? Was it correct? Was it motivating? Did you have the right tools and support to do it? Should we keep what we did or are chimes of change ringing? Despite the scientific and obsessed aspect of measuring OKR, there is also room for the heart. This continuous performance management is considered implemented with an instrument called CFR, an acronym for conversations, feedback, and recognition.

By conversation, we talk about the two-way communication between the manager and the direct subordinate. Feedback is that same interaction but with the peer network, or others interested in the work of the individual. Recognition is a system that allows expressions of appreciation for special contributions. Adobe, for example, conducts these sessions every six weeks, maximum. These three mechanisms exist in most large organizations, although they are called differently.

Again, the *how* makes the difference, since CFR success lies in the authenticity of each of its components. In an environment of mistrust, it will not work. The organization must realize its dishonesty. If you send reprimands to your employees for being late, while being fully aware that they leave at an hour well after what your contract marks, the company is taking an asymmetric position regarding the "social contract" it has with its human resources—and breaching the legal contract. Employees recognize this immediately and an atmosphere of suspicion and mutual cynicism is generated. The same is the case with some companies that have implemented skin-deep systems similar to CFR, for example, to give credit. Modern recognition is based on performance and it is horizontal, that is, promoted among peers. When they celebrate the achievements of their coworkers, a habit of gratitude is born. When it is spurred on by corporate interest, culture suffers.

In summary, OKRs are clear and explicit containers for leaders' priorities and ideas. The subsequent management of these containers is the key. CFR help to ensure that those priorities and ideas are conveyed. But objectives cannot be achieved without a means: the culture of an organization. An OKR culture is a responsible culture. You don't push toward a goal just because the boss gave you an order—you do it because each OKR is important to the company, because you feel respected and protected and because your colleagues count on you.

FILTERING AND KICKSTARTING

We have already designed a system to categorize ideas and map our processes and customer journeys. We know where it hurts the client and ourselves, and we have a clear tree of objectives from the widest to the most concrete for each of our workers. It is time to assign people to projects and get started.

Well, how do we choose where to start? It is important to spend a few weeks on all of the above (mapping, discovering pain points, etc.) with some executives and trusted clients involved and excited about working on analyzing unmet needs. Do a study of what the competition is doing, it may give clues. At this point it is not imperative to come up with a flawless portfolio or do a thorough analysis—bad ideas will soon fall into the flow. Relax. Starting is the priority.

We should hold a session with all the managers involved, preferably facilitated by someone external or neutral, to analyze these weeks' conclusions. During this session, one should not only extract individual projects but begin to glimpse complete areas of growth or need. For example, "It seems like everyone envisions a problem with the purchasing flow," or "It would be lovely to start having dashboards for sales executives." Or maybe someone foresees an opportunity for international expansion in Africa with some slight adjustments to the current product. Perhaps several clients want to extend the range of colors and materials of various textiles that we produce. Occasionally, the benefit is not only economic. In others, the economic size of the opportunity is so blindingly large that we forget that it simply isn't feasible to do so. As a rule of thumb, it is good to lay out the opportunities we choose not only by their financials but for three additional features: it is something that a customer needs; it can be solved by through technology, a product or service that we know how to obtain or develop; and it is something we are in an advantageous position to obtain compared to the competition. Sometimes we have interesting ideas but with a huge learning curve. Sometimes we imagine that the idea has a demand, but it does not, or the competition would duplicate it so easily it that is not worthy to do it.

Defining and writing a filtering system will come handy: what we will do, what we will not do (the most important), and what we will take into consideration to discuss internally. The better defined our filters are, the more automatic decisions can be made without organizing committees and more meetings. In the real world, the first innovation projects will all be continuous improvements. People understand their business, they can cite a good number of pain points without a prior audit. The ideas that emerge are, in most cases, palliative for the immediate. It is important not to lose sight of the future and that current operations have an expiration date. As a rule of thumb, starting with 80% continuous improvement versus 20% radical innovation is a solution for those who don't make up their minds. Or 70%, 20%, 10% if we use the three-horizon system.

An innovation system should behave like a startup gathering. A sponsoring committee should be formed, in the style of those of television shows such as *Shark Tank*. The budget is limited, so the committee must be made up of a group of high-level leaders with sufficient autonomy to make decisions about starting, stopping, or redirecting projects. There should be individuals outside the steering committee and at the same time that not all of the committee are exactly replicated in it. If this is done, the daily operation will end up entering into discussions about short-term projects. It is important to have a pre-assigned budget so that it is not necessary to request funds after completing the filtering and voting process—a common bag for innovation projects. If not, each project must have an executive sponsor.

Depending on the ambition of the initiative and the budget size, the continuous improvement projects should also go through this filtering system. Another option is to have them on their own flux and compete with each other. In any case, they should be linked to the current strategy and be managed mainly within the organizational structure of the company. These are projects that are expected to offer rapid and substantial returns shortly and that need to be financed at scale, so they should be carried out inside the organization. Having well-designed value chain mechanisms for our ideas is a must. The most radical projects, on the other hand, tend to be better off aside from the daily hustle and bustle of the headquarters. Once we have located a suitable group of people, we may be interested in defining workgroups dedicated exclusively to working on these ideas. A best practice is to separate these in buildings and organizational charts completely apart, such as Fjord in the case of Accenture or Space10 in IKEA.

There are several ways to filter projects. For example, the agile scaling methodologies that we will see later propose voting systems and cut notes so that the projects with higher scores go ahead while there is remaining budget. When the last one that fits in the bag is reached, the tap is turned off. When you have very few projects, another simpler way is by simple majority voting. The main venture capital funds do not follow quarterly or annual budget cycles like other companies. When a startup is successfully consuming stages, completes a milestone, or has a problem—and is backed by its fund—it just tries to get more investment, not wait until next year.

What is next? When we have been working under an innovation scheme for some months, we can consider creating specialized functions: someone in charge of generating alliances with universities or research centers; an expert in charge of customer experience or design; an explorer who watches over the market. These systems often uncover problems in the corporate culture, which we will deal with later. Someone from Human Resources specifically working on this project may be interesting. With a modest portfolio, it is possible to use collaborative documents like Google Sheets to maintain it. From a certain point on, you may want to explore specific innovation management software. There are many solutions, such as Planbox or Crowdicity. Some are free. Another specialized software, adapted to the agile methodologies is the Australian company Atlassian, which has a tool for project management called Jira. Very popular, Jira has a collaborative environment like a wiki designed for managing and sharing of knowledge, called Confluence.

Above all, the important thing is to start. Most of the innovations come from dedicated people who work hard to solve a problem and who do not follow a specific method. All methods must adapt to the reality of organizations. The human factor is insurmountable.

Before launching, one last tip: the protocol described so far is nothing more than an orthodox way of doing things. In the real world, it is valid to try to start some pilots without having well-described objectives or without a complete process map. The important thing is to start. Start small. Simply, if you are going to do it, do it seriously.

What do I mean by "seriously?" Think about this: most startups fail to survive the first year. They are started by extremely intelligent people who are dedicated to their project, under pressure for raising a little venture capital, and with all the energy and desire to do things. If they don't manage to thrive, why should part-time employees get it? Assigning part-time people to innovation projects is a typical and fatal mistake of traditional organizations. Even a small pilot needs people 100% involved. It doesn't matter too much who—although another common trap is to place people without any decision-making power. If your organization is beginning to be interested in innovation, everyone is a novice. Never mind. It is necessary to start. But let it be with dedicated individuals, with degrees of freedom to make decisions and away from a harmful "culture of error." Make it clear that there will be no penalties for purging failed projects. On the contrary: many organizations celebrate them because it gives them information on what should not have been done before it is too late.

COMMON WALLS

7

"I can't believe what you say, because I see what you do."
—James Baldwin, 1924–1987

VERTICAL PYRAMIDS

Many managers fall into this productivity trap: more people means more progress. But bigger doesn't mean better in organizing a team—it is often the opposite. Large institutions long to imitate startups in their ability to adapt, but this is impossible for various reasons. One is obvious: the company's size. But there is something subtler, which has to do with hierarchy and verticality.

Most organizations understand career paths as lengthy walks to heaven, climbing corporate ladders populated with bombastic names. The usual trajectory of a professional career is one in which a trainee becomes an analyst or associate, then a manager or partner, perhaps a vice president, during several years of service. Some companies even have timed career paths, namely consulting firms: you know where exactly you should be in three years, four months and seven hours. This is terrible for younger generations and an anxiety machine for talented people, including those in whom the word "ambition" constantly leaves their mouths. At the other extreme, some startups bet on horizontal ambition. Employees are encouraged to dig deeper, become experts, vary their projects, discover fresh ideas, broaden their knowledge, and improve personally, without being accompanied by a new label. We have lost the veneration for crafts, but there are those who still long to be master craftsmen. In other words: developers who want to be great developers, not managers of development teams with dozens of individuals in charge. This way of thinking defines startups and fuels their ability to react.

Some startups also grow, and with growing they face the same challenges. They can verify in their own flesh that generating administrative layers is not useful, because it draws away from the benefits of running a startup. These enterprises are the ones that, over the years, have devised self-management models that today many huge companies try to adopt, without taking into account that these systems were created by the need to manage themselves under different preferences towards growth from those of a large organization. A well-known example is the Swedish Spotify. What happens then? By wanting to embrace other's methods, but not another way of thinking, big corporations fail.

It has been empirically proved that members of a small team are more productive as a unit than those of larger teams. Although each additional individual increases the total productivity of the team as a whole, research shows that the differential in their contribution does so at a decreasing rate. Said more clearly: the fourteenth member has a lower relative input than the fifth. Less and less value is added per person. If you are an economist, this will remind you of the Law of Diminishing Marginal Utility. In 1913, Maximilien Ringelmann asked some volunteers to perform the simple task of pulling a rope. He found that when only one person pulled, he or she made a total effort, but as more people contributed and the responsibility waned, their individual effort ceased. Bibb Latané conducted a similar study in which participants with their eyes and ears covered were requested to shout as loudly as possible. The volunteers made less noise in a group than when they shouted alone. In psychology, this phenomenon is known as "social loafing."

The English anthropologist Robin Durban limits the number of people who can fully interact within a system to approximately 150. From his studies comes the limit with which some agile methodologies draw groups of people with a certain dedication. For example, the tribes or chapters in the Spotify model, or the "trains" (ARTs) in the SAFe framework, which we will describe in the next section. Dunbar relates this limitation to the size of the cerebral neocortex and its processing capacity and adds that, in the case of social relationships, these, in turn, have levels of closeness and quality within this group: we dedicate 40% of our time to five people, 20% to ten more people. In this way, 60% of our leisure time is invested in a close group of only fifteen people. By applying this to workgroups, we cannot spend quality time in groups larger than fifteen, and we will have no connection to some people in groups exceeding 150. Other anthropologists, such as Peter Killworth, speak of a larger number, among 231 and 290. Whatever it is, everything points towards generating smaller groups to allow direct interactions.

Pyramidal hierarchies struggle in creating effective connections and communication within them. You have probably played "Chinese whispers" or the "broken phone" game as a child: a group of people line up and pass on a message, spoken or signed, without the next link in the chain being able to observe previous conversations. At the end of the array, the final message looks nothing like the original.

The same occurs in business communications. After a while, the message breaks down. In some places, skipping the hierarchy is explicitly prohibited, so direct talks with individuals at a hierarchical distance of two or more levels are prevented. The quality of the message, beyond a certain level, irretrievably disappears. And it seems that this works not only with speech but also with memory. At Northwestern University in the United States, an experiment was conducted[1] with 12 volunteers. It went like this: on the first day, the volunteers learned the position of 180 objects placed on a grid on a computer screen—a specific box for each object. The next day, the objects appeared in a different order. They were asked to rearrange them, placing the same objects in the same place on the grid as the previous day, and on the third day they were given the same directions. The result was that the final item arrangement on the third day was more similar to that presented at the beginning of the second day than to the original. That is, even though participants were fully aware that on the second day the objects were disordered, and despite their last memory on the third day was the rearrangement they had carried out themselves, the error in the arrangement of the second day had influenced and changed their memories. This study suggests that our brain is as responsible as our language for the degradation of the message over time.

Thomas Allen plotted a decreasing exponential function in the 1970s[2] for the frequency of conversations that people have according to the distance in which they are placed in an office. From eight meters away, it resulted that the probability fell below 10%. Even though we have many more channels of communication today than we did in the 1970s, face-to-face talks still matter. "If a team can eat more than two pizzas, it's too big," says Amazon CTO Werner Vogels.

The number of possible communications between members of a group is $n * (n-1)/2$. This is a well-known formula in most project management certifications, such as the PMP. It is called Metcalfe's Law. If you studied mathematics, you will have noticed that it generates a quadratic sequence: 1, 3, 6, 10, 15, 21, etc. This law further explains the concept of network externalities that we mentioned pages ago: the value of a communication network - interconnections between people - increases quadratically with the number of participants. The more individuals use an app, its value grows exponentially, but it is a negative thing when it comes to keeping all members informed: a group of only 50 people has 1,225 potential communication channels. Many experts in organizational psychology point to this problem when managing links between members; the cost of coordination proliferates with each additional member.

To cope with this growing number of connections, the preferred form of communication is to organize meetings. We hate meetings. And yet, we should not blame them entirely—much of the responsibility lies with ourselves. We don't know how to have successful meetings. Our meetings are usually organized with dozens of people, which damages the minimalist principle we have just explained. Consequently, our work gatherings become tedious. Too many people do not even intervene, we don't follow the script, and the reason for the meeting is not respected. Above all, we do not do what we are supposed to in a meeting: make decisions. In many cases, decisions will depend on hierarchies beyond those participating in the meeting. Why meet, then? Each meeting should have a very clear agenda: what will we cover, in what order, and how many minutes spent for each task. We often book meetings for an hour, when people can't hold attention for so long in a row—the same goes for presentations. They should only last 25 minutes. Why 25 and not 30? Setting this time allows groups to have consecutive meetings and not be late for the next one. During those 5 minutes the transition is made and you even have time to check the mail. The organizer should share the agenda before the meeting so everybody can get ready. This is especially relevant for the more introverted members, who only dare to speak when they have been able to meditate long enough on what they are going to say. Having an agenda also makes it easy for the moderator to control the meeting, helps employees stay focused, and even lets them decide for themselves whether to attend.

Here are the two significant problems of vertical and pyramid hierarchies: they produce irrelevant meetings and generate endless flows of communication, degrading quality along the way.

To solve this, some radical proposals break with hierarchical structures. In 2015, the company Zappos became famous for proposing two options to its employees: fully adopt an alternative management system, called *Holocracy*—which had been partially implemented for a year—or receive a substantial liquidation and leave the company. 86% of employees remained.[3] At that point, Zappos had 1,500 employees and had been purchased by Amazon six years earlier. A "holon" is something that is both whole and part. For example, a fractal is a fruit tree that contains seeds, but in turn, those seeds contain the components of the tree. In the same way, holocracy seeks to organize personnel by concentric roles and distribute authority. The idea is that everyone can work on several projects while changing roles. There are no job descriptions, but each worker assumes a specific role with clear responsibilities, which may vary depending on the team with which they collaborate. In a holocracy, there are supposedly no titles, no bosses, and no hierarchy, although some critics point out that the flow of decision flows only from concentric circles outward. In any case, we do not witness the hierarchical rigidity of the tree structures to which we are accustomed.

Structure is one of the most controversial issues in organizational transformations, if not the most. The resistance to this change is usually huge. Organizations that need to digitally transform often have templates that are old enough that they are not interested in any changes to the status quo. We are animals of custom. In some cases, it becomes a matter of survival. Even it can be difficult for management to want to adopt new ideas. We are afraid; we risk too much. We prefer to innovate only after others have tried and tested it first.

UNREALISTIC EXPECTATIONS

Human beings are fascinated by stories. We know that Isaac Newton discovered gravity when an apple fell on his head at the foot of a tree—except that it wasn't. The apple story probably never occurred. An apple is the omnipresent cursed symbol: in Eden, in the twelve tests of Hercules, in the tale of Snow White, in the Trojan Wars. Newton was the first to describe with precision and mathematical language the movement of bodies, work that took him approximately two decades, until the publication in 1687 of *The Mathematical Principles of Natural Philosophy*, better known as the *Principia*. He was accused of plagiarism by Robert Hooke, who had published an incipient paper on the movement of bodies and the Earth in 1666, the same year that Newton left Cambridge and began working on the matter. Newton himself admits in his prologue that other authors had noticed the link between free fall and the square of time some 40 years before his book. Human understanding of gravity is very old, although the relationship between time and height with an acceleration constant had not been described with mathematical formulas before. Egyptian pyramids and Greek temples would not be possible without an understanding, at least intuitively, of the phenomenon by previous generations. There are thousands of examples like Newton's. Archimedes likely did not run naked shouting "eureka" through the streets of Athens just because this anecdote was recounted by Vitruvius two hundred years later.

Why does this happen? Because we humans think in "story format."[4] If the story is beautiful enough, its truthfulness takes a back seat.[5] Not only that: we are anchored to dozens of different cognitive biases, such as those described by Kahneman and Tversky. For example, we are a conservative species. The United States continues to use the imperial metric system by tradition, even though in its home country, Great Britain, they use it less and less. Only Liberia and Burma do it. What advantage does this have in a globalized world? None, save for the custom.

Because of these biases, the best ideas don't always succeed. The realm of innovation is not meritocratic, at least not always. Great ideas tend to be strongly resisted until they prevail. The factors are multiple. It can be cultural, by tradition, political, economic. Short-term thinking and the personal agenda of executives frequently come on the scene. In the same way, innovating just for the sake of it is not the answer. It has an obverse and a reverse: the inventors of the plane did not think of military fighters. Henry Ford did not foresee the thousands of deaths in traffic. When we look back, we homogenize and tend to sweeten. Sometimes there are good ideas that your time has not yet reached and other times the opposite: it has happened, but nobody has noticed. The keyboard we use, QWERTY in the case of Spanish speakers (with slight modifications between countries—for example, the French have the letter A in the position in which the Spanish have the Q), was an excellent idea in the beginning. The idea was to mix the most common letters with others in key positions, the central ones, so it slows down one's writing. Does this make sense? Yes, when you live in the last century and most typists are expert scribes capable of reaching such a speed that they lock the rods together. Does this make sense today? It does not seem so. In fact, there are alternatives, such as the Dvorak keyboard, which concentrates the keys that have the highest incidence, specifically 70%, in the center line. The center line of the QWERTY keyboard is made up of keys that in Spanish have only a 32% probability of occurrence.

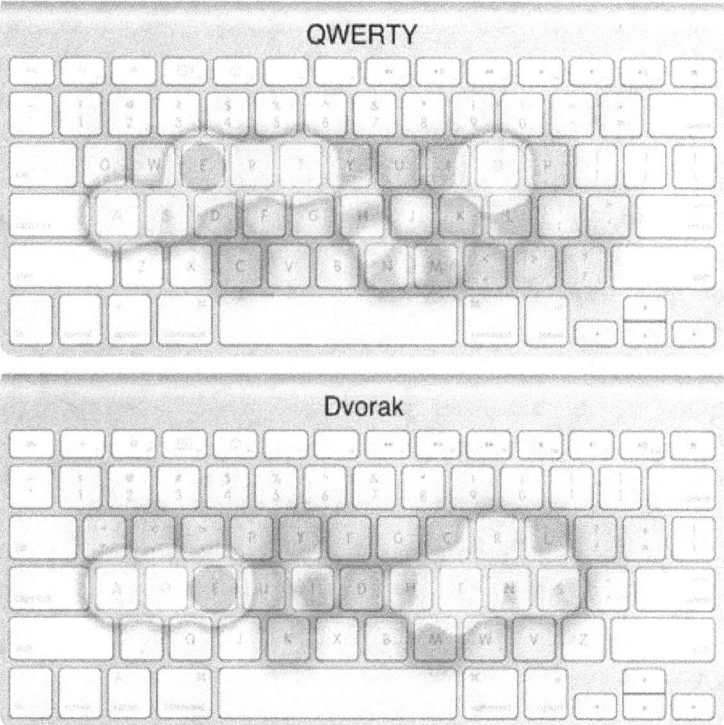

Figure 16: Spanish typing frequency of the QWERTY and Dvorak keyboards. *Credit: Xakata.*

Richard Feynman had lucid moments when he shaved in the morning. At night, some people tend to feel creative in the last minutes before falling asleep.[6] Feynman used to come up with things related to his job (Physics), indicating that we work within our expertise fields. No new ideas for a crime novel plot came to him—he never wrote one, at least. Discoveries and inventions develop from past and present mental work and, strangely it may seem, often during times of rest[7]. Good psychological habits lead to innovation.

Certain fables associate innovation with a mysterious halo. It seems reserved for startups in Silicon Valley or large companies investing unachievable figures for the rest of us. Yes, some started from a garage—not all that we think—but the myth seems to imply that they were destined to succeed. They were well-deserved creditors for Hollywood movies, led by quirky and eccentric geniuses. Many "eureka" moments do exist, but they come more often from errors and accidents than from epiphanies. Teabags were first used as simple containers and did not take part in preparing the drink until the 20th century. Microwaves arose from an unexpected discharge from a radar system developed by Percy Spencer during World War II that melted the chocolate bar in his pocket.

The myth of innovation as radiance out of nowhere, fueled by stories of bohemian and almost mystical inventors, is dangerous. Thinking that the best ideas always succeed, or that there are no irrational obstacles, can lead us to unrealistic expectations. Innovating is more boring than it sounds; they are tedious and mundane (and sometimes unfair) although the discovery of new methods, services, or products can be fascinating. When you hear a story about a sudden spark of genius, it's helpful to ask yourself: How many hours did the creator spend working, how many people made up his team, what was the inspiration-work ratio? Another aspect to investigate is how many attempts, tests, and errors occurred even after having that flash of insight. The formidable insights that come out polished on the first try only occur in fiction. "When inspiration strikes, let it find me working," said Picasso. Well, having a method doesn't guarantee that one will become Picasso, but it helps. The challenge with creative work is the many factors that are out of control. Managing an innovation portfolio is like managing a stock portfolio, where a range of risk is assumed through multiple ideas or investments, but where all can end in losses. You can do everything right and fail. But it is better to start doing it than doing nothing.

Companies are not very original when it comes to competing. When they are in trouble, they quickly turn to lowering the price, or offering more for the same, or creating a temporary promotion, or generating more advertising—but little more. This is because the people in charge of the products are usually the same people who are in charge of operating them, and there is no one dedicated to thinking and imagining alternative ways of selling. The possibilities for innovation are almost endless. To differentiate oneself is not as precise as myth makes us believe. Experience shows that the most innovative firms combine several types of innovation at the same time. Larry Keeley conducted a study in 2011, dividing a group of North American companies into two. The first set was the average innovators, which combined 1.8 types of innovation as defined by them, with only 16% using three or more. The other group, the largest innovators, used 3.6 types of innovation on average, with 71% using at least three types in combination. Keeley then looked at the behavior of the stock on the S&P500 between 2007 and 2011. Out of a ratio of 100 at the start of 2007, businesses that used between 3 and 4 had a 50% increase in share performance, and those using 5 or more doubled it. In contrast, indexed companies without registered innovations were losing value. Another interesting conclusion is that product-based innovation is less and less used to the detriment of combinations of others, such as processes, networks, or adhered service.

Figure 17: innovative companies' performance. *Credit: Ten Types of Innovation (Keeley et al., 2010)*

Starbucks wanted to be a gourmet coffee bistro first. Today, it is more recognized for its space than for its product. Their coffee shops became temporary offices where meetings, job interviews, or teleconferences take place and it is a refuge for writers, freelancers, and anyone who urgently needs Wi-Fi. When organizations see that they are not selling the way they intended, they repeatedly ask themselves whether it is necessary to increase the product quality, lower the price, etc. Any of those strategies might build up sales, but you can't be sure either. The approach is often unscientific and highly dependent on luck. Asking the right questions is important. Instead of: "How can I get more people to buy my product?", they should ask: "What role are my products or services playing?" In the case of Starbucks, its clientele is not limited to coffee connoisseurs. All products and services serve a purpose, but that purpose, as we saw before, differs depending on the context. Clothes serve to cover. Furthermore, they can serve to cover in a certain, fashionable way, to improve our appearance. We might need that improvement to be repeatable: "give a good image every day in the company" or eventual and urgent: "the airline has lost my suitcase, get whatever for the meeting this afternoon, but make it elegant." The need could be rather insignificant: "I need some wool socks to walk around the house," or something critical: "the dress for the graduation party." All products and services play a role—try to understand what it is. Even behind the most rational decisions, there is an emotional component. The rationale "we were hired for the new industrial plant" may contain, behind the scenes, a "Natalia, the project manager, might get a promotion with this work and is willing to commit the next year's budget." As we saw in Kahneman's and Tversky's experiments, people are terrified of the possibility of losing. When we seek to innovate, we often think of perfectly rational and functional aspects, and psychological aspects are discarded.

Problems can become opportunities if you look to solve them. Diseases led us to create sewage systems, vaccines, and medicines. Look for solutions to each obstacle, innovation starts with understanding the key but not obvious elements. When selling a door, its material may not be the critical factor, unless it poses trouble per se, because of its cost, its weight, etc. What role does a door play? Try to find an original way to get in or to keep strangers out—those factors may reveal key elements. If you start there, ideas will appear soon. If you never find the keys in your bag when you get home, you may be the one to invent a voice recognition door opening system, as there are already in some call centers. A righteous attitude helps innovation. The creative problem-solving technique involves combining two or more ideas or concepts to see what new product results from there—in other words: getting from point A to point B in a way that no one had thought of before. Crucially, you assume there will be a solution, so your mind works on something it hopes for. No brain works hopelessly. For the same reasons, fun and diversity help innovation. When you entertain yourself, your mind works without realizing it, and different teams can come up with brilliant solutions by contrasting ideas. Remember that there is no single form of intelligence, and the IQ calculated in tests is often related to analytical-mathematical intelligence only, which does not necessarily correlate with creativity. Innovating is tedious in the literary sense, but it must be fun in the day-to-day of the people who are involved in the process.

We have already seen the need to document the interactions that clients have with an organization: customer journeys. When the right questions have been asked and we understand the process by which customers arrive at our products, the rest is just a cleaning job. We must then remove obstacles, remedy frustrations, and invent a better experience than we have today. Several factors can give us clues about the impact of an idea or the ease that we will have to implement it:

1) What value does it have regarding what already exists? If it is something intended for a customer, this opinion must be his or hers. In the case of a new process or internal system, we can check it with the users themselves. We can even ask ourselves the following question: will the results be visible? Showy innovation is always much easier to scale in an organization.

2) Can we easily test it? We must always remember the importance of prototyping. If we can't make a prototype to show potential customers, we may not be able to answer the first question with total accuracy.

3) How much effort does it entail? Even having a positive answer to the first question, we can find that the cost of implementation is much higher than its subsequent benefits. An internal or external business case is a good idea, although it is important not to fall into a trap. Sometimes the same business cases are used as an excuse not to change the status quo.

4) Is it the right time for this idea? This question is tough to approach. The iPad was not the first tablet. Other pads existed, but they did not receive the same acceptance, although they contained many of the same functionalities: touch screen, applications, etc. It was too early and its use too complex. We must assess whether the complexity of our idea will be too much for the target market.

THE ELEPHANT IN THE ROOM

What to do with culture? It is intangible, invisible, inaudible. Surveys recognize, however, that culture is as or more important than strategies or operational models. This vision of its importance applies worldwide and in any industry.

Culture exists. Although we cannot see it, we smell it as soon as we pass through the door. As in a date, where first impressions are crucial, those of an organization say a lot about it. In a few minutes, we will have learned much of how it works and what it is: what type of furniture do they have, what colors, what kind of spaces they have chosen as an office, and what type of lightning. Are there open spaces, or do they use cubicles instead? Are there offices, or does everyone have a similar space? Certain woods are signs of status, and it is not uncommon to find solid mahogany and teak in many older corporate buildings, along with shabby furnishings and fixtures, such as framed photographs or diplomas hanging on the walls. The White House resolute desk, a regular visitor photography spot, is made from the wood of an abandoned ship that Queen Victoria gifted to the United States in 1880. Although smoking is banned in most countries, one can smell the cigar aroma of such an environment. All this is in contrast to the Scandinavian-inspired minimalist designs that replace solid woods with agglomerates and synthetics such as melanin, formica or PVC, or even metals, giving an image of an industrial factory. Out there, in the millennial world, they are always chasing for spaces with free food and amenities. The eye captures all this; we understand and internalize it. It is true, we recognize a company as soon as we enter the door, even if it is Sunday and there is no one working.

Thus, acculturation[6] is achieved from the first moment.

Is an organization's cultural change even possible? The same concept of cultural change in adults is scientifically debatable: to what extent is an adult capable of changing the way they act? There are various psychological studies on behavior changes of migrants in host countries. Steven J. Heine, the author of Cultural Psychology, wrote an interesting paper[8] in 2012 together with two of his students, where he measured the acculturation level of immigrants from Hong Kong to Canada. Not surprisingly, they discovered that from the age of 15, the ability to acquire the new culture decreased, and especially from the age of 25. Childhood is our only homeland. It seems difficult for post-university workers with over 4 or 5 years of experience to easily win radically different habits. But in the same way that we take off our shoes before entering someone's house in another country, we can accustom our bodies to pick up other behaviors based on what we see around us. Other studies on the influence of acculturation on people's professional behavior show that deep ethical values change in adulthood after adapting to a new culture, as studied by Jaffe, Kishnirovich, and Tsimerman and published[9] by *Journal of Business Ethics* in November 2015.

---

[6] The process by which a person or a group of people acquire a new culture.

It is difficult to change, but possible.

Business culture is a complex animal. Cultural habits are embedded in people's networks, reinforced and strengthened over time. Most networks of people are characterized by their culture, and in large enough organizations, there are several of these networks—departments, areas, groups of friends, the oldest workers, etc. This fact indicates that there is no single corporate culture. That is why we can understand it not as a discontinuous element, but as a series of layers that shape certain behaviors. Any employee who has ever visited a branch in another city will attest to how things are done slightly differently there. In the IT department, it is usually different from Finance; in turn, all IT departments are different. Even within the same department, a group of innovators and those of the old school may coexist. Each threshold crossed leads us to a new planet. The national and regional culture itself influences with different intensities. But the culture within similar teams can also feel very different. Who leads the discussions? Do people jump into ideas, wait until the end, or use email to give suggestions?

People enter and leave their companies frequently. On average they remain a few years, and this average is increasingly shortened with each new generation. But far from being renewed, the culture is stubborn. Newbies often adapt quickly to bad habits. Corporate culture is often an example of a self-fulfilling prophecy: the people who stay the longest are precisely the people closest to the prevailing culture and not vice versa. So, as if it were a symbiote or a parasite, culture struggles to stay. And usually, it does.

Rather than tackle the rocky problem of transformation, for some, culture begins with recruitment. At firms like Amazon, recruiting is a global responsibility and each employee is expected to collaborate in finding the best talent for the company. Once inside, a correct communication of the mission and vision helps. It helps to have present leadership that doesn't lock into its glass tower and debate side by side with each team member. What is sought is assimilation or integration that is not followed, as it sometimes happens, by separation. Nations have powerful governments that have a monopoly on violence, make laws, and can enforce them. Although organizations have governing bodies and regulations, it is much easier for anyone to transform organizations than countries, even in periods of high unemployment. Therefore, the coercive layer of change management tools is frequently discouraged.

Companies operate in many ways similar to nations, so we can apply the sociological and anthropological theories made for larger ethnicities and societies on a smaller scale in a corporate building. For example, new hires routinely experience the so-called *immigrant paradox*: recent immigrants outperform established and nationals by some parameters, despite the hard barriers they must overcome. Recent hires enter with an invigorating force and culture until they assimilate and normalize. The average permanence in a company not only tells us something about its culture, but it also informs us about the performance that we can expect on average from its employees. Immigrants bring a backpack with enthusiasm and desire—new employees, too.

Since companies are equal to nations in some facets, we can learn from the Four Types of Acculturation theory, sometimes named the Fourfold Model,[10] and apply it to private organizations. According to this theory, there are four ways of transferring culture, also recognized in four different types of organizational culture:

- Assimilation, which occurs when new hires adopt the new culture. In the case of nations, this is frequently enforced by governments.
- Separation, which occurs when individuals reject the new culture to maintain itself. This occurs both in

wrong hires and false promises about how an organization works.
- Integration, which occurs when individuals fail to adapt to the new culture while still maintaining itself.
- Marginalization, when the individual rejects both their culture and the host.

If you have paid attention, you will have observed that the individuals adopt or reject, but they are not able to transfer their culture to the recipient nation, except in cases of sudden and numerous migrations, for example the Italian influence in Argentina. But that is difficult to achieve by hiring new people when the organization already brings enough historical inertia. We need something else. What is sought is assimilation or integration without separation.

In any transformation initiative, management's job is to discover how to take advantage of the positive cultural traits of people affected by the transition to build momentum and create lasting change. Companies that succeed in doing so by adopting a 'culture-driven' approach to change substantially increase the speed, success, and sustainability of their transformation initiatives.

Organizations that have managed to capture or design these traces use specific acculturation techniques when receiving new employees or members. Several have training and induction programs in the first days or weeks. Some assign a personal mentor to help with simple tasks like recognizing where the printer is or how to request a trip. Larry O'Toole runs all his new employees around Harvard Stadium. Others teach the language of the environment, ranging from industry jargon and acronyms to meetings or room names. Many people are embarrassed at first to ask what an "XGB" is under the assumption that they are ignoring something obvious, when in fact it is an acronym that I just made up while writing. It is critical to implement this type of initiative in order to facilitate the assimilation of the organization's own knowledge in individuals who have just arrived. The cultural fit is so vital for so many companies, especially in the Anglo-Saxon world, that many declare they hire more by attitude than by aptitude. Skills can be trained, attitude cannot. Conversely, a firm with a toxic culture will annihilate the attitude of any new hire.

Why should all of this matter to us when we talk about innovation? Because the reason why organizations exist, beyond making money, *matters to people*. Behavior that acts both through an abstract entity (the company) or an individual (its employees) *matters to people*. And, as we have seen, the attitude towards experimentation, security, the good environment, or the acceptance of error is decisive in order to innovate.

Companies can devise useful maneuvers in the short term to improve their results: vary the price, carry out promotions, advertising, lobbying, etc. This will typically generate transactions that will work once. It does not serve for enduring loyalty or recurrence. The real mirror of inspiration are organizations that don't need tricks to sell their products.

Acculturation in a group is so important that it goes beyond its physical limits to reach its clients. Some studies have analyzed the impact of cultural adaptation on product sales. Customers not only buy a product, but also buy according to the brand and what cultural traits this firm provides. An example is western food companies adapting to the growing religious needs of their local market. Kentucky Fried Chicken decided to launch a halal line for Muslims, and other businesses have kosher lines dedicated to Jews. This type of innovation can only start from people with a certain sensitivity to other types of realities (in this case spiritual).

Every morning, culture eats strategy, Peter Drucker said. Buying technology is sexy, easy. Implementing systems is not so much, but it is always preferable to implanting a new chip in our heads. Whether or not we like it, digital transformation has plenty more to do with cultural transformation than with the adoption of technology. Certain technologies compel discipline and further structure within our data and processes. They reveal problems and discover bottlenecks. But in the last stay, the steppe wolf will come out. If we do not have the necessary cultural talent, it does not matter the investment we make in systems, or in defining our business model.

Changing the culture of an organization is one of the hardest leadership challenges. It comprises an interlocking set of goals, roles, processes, values, communication practices, attitudes, and assumptions. These elements unite and reinforce each other as a system and combine to prevent any attempt to change it. It is a living and pulsating being. This is why single solutions, such as introducing agile methodologies or knowledge management, or the purchase of some software that "make order" in our minds, can progress for a time. But the core of the organizational culture will eventually take over and will revert to the pre-existing state.

Sociology and anthropology teach us that it is possible to change culture through practices and habits. But the personal agendas of the most influential executives in that organization will always resist. If they don't want to change, the culture will never change. A quick search in the literature on change management will show us countless references and articles with the same advice: start from the top. Many CEOs of the best companies usually send emails directly to all their collaborators, giving a personal and close treatment to their conversations, even if they are by electronic channels. People are, moreover, extremely insightful in capturing when there is no correspondence between the culture they are advertising and the one that actually occurs. False messages about the type of organization we are witnessing or the disinterest in changing the status quo by management will quickly be felt to the last extreme of the organizational chart.

Changing a culture, if truly desired, is a large-scale task, and eventually many different tools will have to be put into play. However, the order in which these tools are deployed has a critical impact on the probability of success. Only a few simple steps are fundamental.

First, evaluate the current culture. It might be difficult, as culture is something intangible, but if it is possible to perceive it, it is possible to describe it. It is important to put it on paper to know how much change we need. When I speak of measuring, I am referring to what employees believe to be the organization's current values, not only a limited group of people. The participation of all interested parties must be guaranteed on equal terms. This can be done through a survey and can be used to expand cooperation in receiving ideas, proposals, and solutions.

Second, define the business priorities: growth, customer satisfaction, etc. Quantify your cultural values and think about how they relate to your goals. Do you need proactivity and individualism or obedience and homogeneity? Which two departments do we desire to put to work together? It is imperative to align culture with strategy. Strategy with structure. Structure with culture.

Third, some things ought to be reconsidered: committees, presentations, etc. When I talk about structure, I also mean procedures and tools. Are all necessary? Is it possible to transform an information committee into a scorecard that keeps us informed in real-time? Now we can think about technology: what technology allows us to save unnecessary things that bore people and keep them from what we want them to prioritize?

Fourth, communicate the conclusions of the analysis and show them repeatedly. The number of repetitions required for a message to stall is colossal. Implement the changes, then review everything once again.

This is it. It is useless to go too far talking about culture. All in all, techniques that could be explained here are helpless against the will. Changing culture is a matter of will. Quitting smoking is a matter of will, as is weight loss (except for endocrine complications), too. Studying instead of watching television is another example. Nicotine patches, group therapy, sports, a properly prescribed and followed diet... All this helps, but nothing compares to an iron and irresistible will to do it. Implementing systems makes us more disciplined to work in a certain manner. These systems include hiring external experts, adopting agile methodologies, hiring for attitudes and not for aptitudes. All of this helps tremendously. But nothing compares to determination. Nothing will work if there is no desire to achieve a certain way of being and acting.

Al Pacino sums it up well during the *Scent of a Woman*'s closing speech: "I have come to the crossroads in my life. I always knew what the right path was. Without exception, I knew. But I never took it, you know why? It was too damn hard." Most of the time, we know what to do. We are simply not interested.

THE CUSTOMER AS A MYTHOLOGICAL CREATURE

Technology is fascinating. Software systems are capable of speaking in natural language, understanding each other, even making jokes. There is an application for every need. Sometimes the solution is simpler than that, however: it lies in producing something beautiful, simple, and well-designed. Too often, beauty is circumvented in the corporate world. There, they prefer tedious presentations with infamous designs, overcrowded letters, and too few chromatic or visual indicators that would help to memorize or understand better. We use PowerPoint just like we use Word, except with a horizontal canvas. We live with the feeling that behind an overweighed Excel there is a strict analysis that cannot fail; behind a slide with 500 characters, an impeccable job that relieves the pressure on our shoulders and assures us that what is being proposed is, without a doubt, the right thing to do.

In reality, none of this is true.

Several clues show that functional and beautiful design is essential in all aspects of our lives, not only in product management.

In 1998, a startup called Google published its search engine for the first time. Until then, most of them operated with databases populated by the administrators themselves. If you built your personal website, you needed to "warn" the search engines of the time that your website existed. On my first website, there is still the message: "Register your page in all these search engines with us," then a link to a search engine aggregator. The privacy of the world wide web was another matter. The pioneers in searching, W3Catalog and Aliweb, worked this way—most of the popular search engines during the nineties like Lycos, Altavista, or Infoseek did as well, although each one innovated in some way or another. Yahoo!, founded in 1994, started as a web directory, although it quickly developed its own spider or web crawler. They are robots that, walking through the great web, allow a search engine to "discover" what new pages are appearing without needing to be notified. Google developed neither the first spider nor the first indexing system; the credit goes to JumpStation, who created the system in 1993. They were not even the first to produce a sorting algorithm that would rank web pages within a punctuation-ordered index and rules of importance. This credit goes to Robin Li, who developed RankDex two years before Google launched PageRank, the algorithm that made them famous. RankDex is still used today under the hood of the most used search engine in China, Baidu. Inspired by the academic world, where the most referenced papers and authors take on greater importance, both algorithms associated the quality or importance of a website with the number of links or hyperlinks that pointed to it. Some Google patents related to PageRank refer to Li's previous work. With PageRank, Google exploded, and it's only fair to give it due credit.

But there is one more detail that separated Google from Yahoo! in those incipient times: the design of their home page.

Figure 18: Yahoo! home page, 1999. *Credit: Internet Archive.*

In the same way that specialists recommend limiting the number of words on a PowerPoint slide to less than 10, Google maintained for several years a "non-written policy" that limited the number of words on its welcome page.

Google's was simple, white, and clean as snow—simple to use. It just asked you what you wanted and found it for you without further inconvenience.

Yahoo!'s home page started as a directory containing news, announcements, and other links in addition to its search engine. After 2000 it became increasingly complex, succumbing to its advertisers. In 2004, there were 255 links on Yahoo!'s home page. "It had nothing to do with the user, but what Yahoo! wanted the user to do," Tapan Bhat, vice president at Yahoo!, told the Wall Street Journal in July 2008.[11] Yahoo! tried to refocus. By 2006 there were around 170 links on the home page. In 2007, it was reduced to approximately 140, in 2008 to 120, and in 2009 to 100. It was not enough.

We will never know how much merit is attributable to internal indexing algorithms. And, of course, the internal culture of Yahoo! and Google have their role in this story. But something as simple as the design of their front page had a direct impact on the success of both companies.

Figure 19: Google home page in 1999. *Credit: Internet Archive.*

We find something similar in Steve Job's famous speech[12] at Stanford, in 2005:

> *Reed College at that time offered perhaps the best calligraphy instruction in the country. Throughout the campus every poster, every label on every drawer, was beautifully hand calligraphed. Because I had*

dropped out and didn't have to take the normal classes, I decided to take a calligraphy class to learn how to do this. I learned about serif and sans serif typefaces, about varying the amount of space between different letter combinations, about what makes great typography great. It was beautiful, historical, artistically subtle in a way that science can't capture, and I found it fascinating. (…) If I had never dropped in on that single course in college, the Mac would have never had multiple typefaces or proportionally spaced fonts. And since Windows just copied the Mac, it's likely that no personal computer would have them.

Apple went far beyond typography: it introduced the first graphical interface through windows and icons on January 19, 1983 — inspired by a Xerox preview that never saw the light of day.

Figure 20: Lisa was the first personal computer on the market with a graphic user interface. *Credit: computerhistory.org*

What did Google and Apple have in common? They thought about what the client needed. The importance of design is just one facet that separates organizations that care about their customers and those that treat them like non-existent mythological animals. If digital transformation is a hackneyed term, no less is the expression "customer centricity," or "putting the customer at the center." All businesses claim to do it, but few do. In the forums where decisions are made in many companies, the customer rarely appears: committees, hallways, informal meetings, shareholder conventions, etc.

We have three sources to get information about what the client wants:

- Our own employees, who are in many cases also customers, of course. A correct internal knowledge management system can be very useful, as well as ideation, discovery and design thinking sessions.
- The customers. This is obvious, but few companies involve actual clients in their decision making. Focus groups, interviews, and qualitative surveys, are a wonderful way to understand the customer's needs.
- Data. Nothing speaks more of you than your own actions. We need not have the client in front of us if we can describe their behavior. Our clients generate data, and then we can analyze it.

Learning to listen to the customer will allow us not only to adapt parts of our offer, such as design, but also to radically change the course of our organization towards better seas. Let's see some examples.

What is this company doing to make their core offering compelling? How do their shareholders feel about side projects like Twttr when their primary product line is, besides the excellent design, a total snoozer?

This was one of the first criticisms[13] received by a pilot project, run in parallel to the primary business of the Odeo company in TechCrunch in 2006. The name of the project was Twttr, later renamed Twitter. Those who listened to TechCrunch's advice missed one of the biggest investment opportunities of the decade. As a curiosity, the Twitter pilot test app they use to experiment and test new environments is also called Twttr.[14] Flickr, one of the largest image servers in the world, started as an online game chat room called *Game Neverending* where photos could be shared in real-time. Soon after, they forgot about the game, expanded the capacity and photo upload options, and buried the chat room. Groupon started out as a kind of change.org, with a site called *The Point*, where activists with similar causes connected.

Eric Ries realized that startups managed their meager resources much more efficiently than large organizations and started the Lean movement. Ries's central thesis is that startups do not come from great ideas that make them unicorns overnight but result from a process, rare to find among large companies, which constantly test and learn what customers demand. In other words, startups solve the innovator's dilemma in a simple way: they don't have too many products to improve, so they have developed a unique capacity for disruptive innovation, through constant trial and error. Typically, large organizations do a market study, then generate a strategic plan around an idea and deliver a product. This assumes that you know exactly what the market requires, which is often not true. Startups, however, perform a process of trial and error constantly, and they end up turning into something completely different for which they were conceived in the first place, like Twitter, Flickr, and Groupon.

This phenomenon is old and not tied exclusively to startups. Nokia started by cutting down trees and producing paper. 3M was a mining company, which still retains its original name intact: Minnesota Mining and Manufacturing Company (3M). Nintendo produced card games. They are all old companies that, if they have survived, have been because they have known how to jump out of business, listening to what the customer demanded.

AGILE

8

"I don't play Salsa, I eat it. I add it to spaghetti."
— Tito Puente, 1923–2000

The nineties. We are on the Peruvian national television. Tito Puente, the timbale king, a white-haired septuagenarian with a strong glare in his eyes, answers a question in front of an astonished journalist. She had asked him, a few seconds before, to describe what Salsa —*sauce*, in Spanish— is.

Salsa does not exist, indeed. The term has been controversial for decades before settling in popular speech. It is a commercial word, probably dating from the 1960s and forged in New York, although the first reference is recorded on the radio in Venezuela. This voice brings together rhythms of Cuban, Caribbean, and African origin. There is, for example, the guagancó, which has a defined rhythmic pattern, as well as the rumba, the son montuno, or the chachachá. We could say something similar about flamenco, not a style as such, but a set of so-called palos: alegrías, bulerías, seguiriyas, sevillanas, fandangos, and so on.

Like salsa or flamenco, Agile is a collective noun; an umbrella that brings together methods and techniques with a common underlying philosophy. A methodology is simply a set of rules and steps. What defines Agile is its underground wisdom and the lessons learned. Whether we talk about XP, FDD, Crystal, or even Design Thinking, we all find common features: *they all sound similar.*

It is often said that the agile movement began in 2001, with the publication of a manifesto still accessible today in its original state on the internet.[1] In that document, a group of software developers expressed their opinions and values about how to write code. Some well-known agile methods like Lean or Scrum, however, are older than the manifesto. The exact date does not matter—what matters is that all methods start from a common criticism to the way of doing things, which is typically the method that prevailed during the last century. In recent years, Agile has become fashionable in all industries, and is not limited to software development only. Using agile methodologies is now a kind of obligation in internal projects of companies that want to transform. And it is an inescapable requirement for their suppliers, too.

Have you ever planned something complex and well in advance, and then everything went exactly as you planned? This never really happens; I can't even plan my next day accurately. The most important flaw Agile's pioneers saw in traditional ways of working was the way projects were planned. We typically tackle a problem sequentially: analyzing, planning, designing, implementing, testing, and moving it to production. It is the intuitive way and with which we have been educated at school: doing everything right from the beginning and ensuring there are no mistakes before turning in an assignment. The influence of Taylorism, seen in the first chapter, is perceived. This approach is unreal, for various reasons, especially when we talk about marketable products or anything other people are going to use: we don't know exactly what they want or what they like, nor can we understand their needs by writing their requirements on a piece of paper. And if we manage to write them down, be sure that the needs will change after a short time.

The software engineers signing the manifesto understood this problem. They had followed for years a general process of development called *waterfall*, in which each phase is unleashing the next, in cascade. First, a requirement is made in a formal document. Then, the software architecture is designed, the code is written, and each piece individually tested. Then the entire set is tested. And finally, it is deployed into production. This entire process could take months. The problem comes when, in the middle of the process, the developer made decisions on her own without consulting the client. This may happen either because it was impossible to contact them for feedback, or because the developer's personal approach seemed more reasonable, or because the communication of needs and requirements had not been adequate. Upon the deployment stage—or in the best case, during Q&A—the client received a software that had nothing to do with what they had requested. The opposite effect is possible and even more common: it is the client who changed his mind halfway through the process. By having a linear protocol, that meant going back to the beginning, reworking the technical documents for taking requirements and starting all from scratch.

Figure 21: the original Agile Manifesto. *Credit: agilemanifesto.com*

Over time, it was proven that this process proved to be a disaster, and the way of working needed to be modified. Thus, the pioneers began to draw in their minds the fundamental pillars on which the manifesto was written. There are four, but I like to group them into two pairs that have a certain relationship:

- ✓ **The need to have face-to-face interactions** between clients, developers, and other individuals involved in the project. Communication processes and tools were necessary and useful, but nothing can replace human interaction. Conflicts are best solved in person. By mail or phone chat, people tend to be more hurtful and direct, even unintentionally, because tone and body language are absent in written conversations. If you've ever been in an email thread conversation and tensions have escalated, only to be later resolved in a couple of minutes face-to-face, you know what I am talking about. Do you remember the communication problems explained in the section "Vertical Pyramids"? In the same way, **collaborating with the client is better than negotiating a contract.** Maintaining a human touch (either with the work team or with those who will use the product) is always better than having a written sheet of paper with what to do.

- ✓ Accept that the environment is dynamic and that **flexibility is preferable to following a plan to the letter. Deliver value** (something that works), **rather than worrying about the documentation.** This happened in waterfall methods. When a change came, all documents had to be modified. So, it is better to complete it in a parallel fashion, while using the working code. We avoid documenting each step to advance the project as soon as possible because we know that conditions can vary.

Compared to the waterfall method, agile methodologies propose iterative growth in small increments of value. Technically said: they materialize the Deming cycle—plan, do, check, act—in short portions of time. What characterizes these methodologies is the speed with which they deliver simple versions of the product that complement each other and become increasingly complex in short spaces of time, until completion. This action increases visibility since we don't wait until the last moment to see the final product. In the world of software in particular, and of intangible projects in general, the agile manifesto has been widely accepted, and there are more than a few methodologies that have implemented its values, being Extreme Programming, TDD, FDD, and, above all, Scrum, the more common—Scrum, as we have already mentioned, is older than the manifesto.

This does not mean that Agile is only used for development projects—that is, to create new things. Other management theorists discovered the same mistakes but decided to focus on continuous operation. Two clear examples are Lean Management and Kanban.

Lean Management helps us implement a strategy for large organizations interested in perfecting their continuous improvement processes. It is an old model, which only started to be called lean since the 90s, but comes from the way Toyota worked decades before. Toyota's primary goals were to reduce or eliminate overload, muri in Japanese, inconsistency (mura), and waste (muda). They defined eight types of muda: waste due to overproduction, waiting-time between production processes (i.e., people waiting for the previous in the chain), logistics, processing, stockpiling, movement in the production chain, manufacturing defective products, and underutilized workers. Toyota published an official description of its system in 1992, updated in 1998, and publicly available.[2] The document warns us that it is not an instruction manual on how to do things, but an exposition of their product concept and philosophy. As in the agile manifesto, it describes many preferred values, two in particular: *just-in-time*, or producing only what is needed, when it is demanded; and *jidoka*, which depicts a type of intelligent and assisted automation in which humans limit themselves to quality control operations.

Toyota developed the Kanban system in the 1940s to achieve the "just-in-time" concept. It is conceived as a task scheduling system for manufacturing and spawns from the following hypothesis: delays occur due to an excessive or insufficient supply of resources necessary to complete each process at all times. In other words, we never have the right amount of people at every moment to get the job done. Why? Lack of communication, silos, rigid structures that do not allow the flow of personnel between them. "This employee is in my team and not yours." This issue is just what we referenced in the section "Vertical Pyramids." The solution? Visualize the full flow and rebalance or tackle bottlenecks, helping whoever it takes. In this way, you can be more productive with less waiting time. Interestingly, Kanban has expanded in recent years to all industries, including software development, walking the opposite path to other methodologies, which have gone from software to everything else.

Kanban is therefore a visual way of managing tasks and workflows. It achieves this by employing a board (physical or digital) and a set post-it cards placed on it. Kanban means "sign" or "poster," and the word is made up of two words: 看, "look," and 板, "board." Let us look at how it works. Each card or post-it represents a task on the board and provides information: its name, a brief description, its duration or deadline, and the benefit, importance, or value of that task. The card will be assigned to a member of the group, who will then be responsible for carrying it out before the deadline. The columns on the board allow us to differentiate the different stages of the process. In the headings, there will be the name of the phase—for example, correction of comments and some policies that allow knowing when a task has been completed. For illustration:

- All comments have been corrected by the engineer.
- The document has been sent to the technical manager.

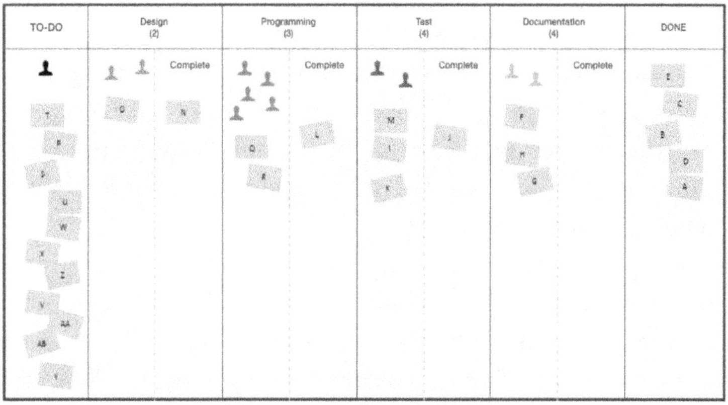

Figure 22: working on a Kanban. *Credit: Management Plaza.*

The card could then go on to the next column that could state: "under review by manager." Each column or stage should have some resources assigned by default—it could be the number of people in a department—and resources assigned at the moment. The numbers will normally match, except in the event of an activity overload or bottlenecks. The cards are pasted under the column headings and dragged to the next column to the right to show where they are in the production cycle. Ideally, we should tailor our dashboards and design them on purpose for each workflow, although sometimes a simplified dashboard divided into just three parts:

to do, in progress, and completed. With such simple tables, we can obtain results.

In figure 22, we see a more elaborate example for a software development process. We have six stages: to be done, in design, programming, testing, documenting, and completed. There are also twenty-eight tasks. The icons show the maximum number of people we have in each department—analysts, developers, testers—and what people are working on each task. It can be seen that there has been a rebalancing of resources: someone is out of the office (pending), two are analyzing, five are programming, two are testing and two are documenting, while the default distribution—written in parentheses—is organized as two designers, three programmers, two Q&A testers, and three people in charge of the documentation. This exchange of assignments is not always possible in all industries but reflects the nimble philosophy of being able to have people with different abilities working on different things.

Flow visualization using columns and cards improves efficiency and avoids overcapacity, as we place limitations on the number of tasks placed at different stages. So, each stage has two columns: "under construction," where the dedicated resources are shown, and "complete." From a "complete" column, it is not possible to jump to the next phase "under construction" until there is the capacity to do so. And therein lies one key to Kanban: not overloading teams. This helps limit the number of tasks in a job in progress, so teams can focus on only those tasks and work faster.

Teams enjoy this system because of its ease of use, visual interface, and ability to see what everyone is working on. It also provides information on the progress of the task and if anything interrupts the project. Now that we have very briefly seen how a dashboard works, you can go back to the example figure and have, in a glance, a good idea of what is happening: percentage of progress, where the bottlenecks are, which departments need help, and at what rate it is delivering. The benefits are countless. A dashboard places the entire process on a single page or screen, making it easy to see who is working on what and where they are in the project cycle — that keeps work in progress with fewer interruptions. The team can then focus on the tasks immediately required, thereby increasing the efficiency of the entire flow. Project managers can assign tasks when a team member is idle, and team members always have work to do. When a task crashes for any reason, a special column is enabled (not pictured). In summary: this process facilitates the smooth movement of work, avoiding delays and overloading tasks.

The Kanban board is part of a larger methodology that needs to be studied in detail. It not only helps visualize the workflow, but it also serves to manage the value chain or process of any production system, from the supplier to the final customer and all points between them. Kanban cannot be an isolated tool, but must constantly monitor its process. You should always be looking for improvements to increase efficiency and keep resources balanced with production demands—*continuous improvement.* I feel that sometimes Kanban is a wasted tool. I have rarely seen boards more complex than those with three columns. I would like to see full departments implementing their flows or, better yet, interdepartmental processes. It is there where its true potential is exploited and where its virtues are best used.

BUSINESS AGILITY

Why do agile adoptions fail in many organizations? I think we can sum the reasons up in two grand families: executives wanting to maintain control at all costs, and the coexistence of agility with traditional practices. Let us have a look at both.

In the chapter "Common Walls," I insisted on believing in what we are doing, on the need for the governing body to be convinced and avoid methods that do not comply with our way of thinking. It was for a good reason. What exactly do I mean? Let us take a small, concrete example: Gantt charts, a classic of project management.

There are two chief reasons a Gantt chart contribution to your project is limited.

The first one is its age; they were invented in 1910. They are, therefore, a pre-World War I invention—in fact, they became popular during that time. Not for the simple fact of being old are they disposable, of course, but from 1910 until today, some important things have changed, such as the speed at which events happen and at which we communicate. Remember the fourth value of the Agile Manifesto: "respond to change by following a plan." Plans change very quickly, and the Gantt charts, which come from a time when everything moved slowly, are pie-in-the-sky in the long term.

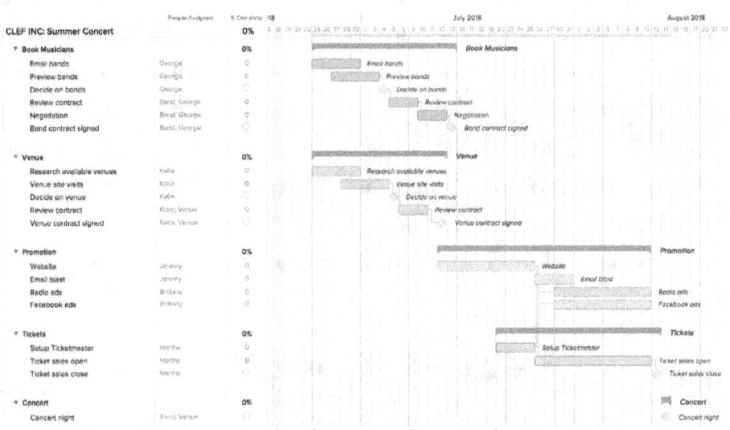

Figure 23: An example of a Gantt chart. *Credit: TeamGantt.com*

The second reason has to do with the way projects are managed. Projects are, in fact, *not* managed. People are. In Microsoft Project, for example, people are just another resource. Certain questions are not answered: What is everyone on your team working on? Who is busy and who is not? That is why systems like Kanban visualize each task, of each person, in each instant of time. This does not happen with Gantt charts.

What the agile mindset is trying to instill is clear and concise: set short-term goals and finish small chunks of work in limited time frames; involve the client from the beginning; promote face to face, openness, empowerment of teams; being willing to fail, not seeking to be perfect the first time; make it simple. You can use a Gantt chart as a support tool. But if you are truly concerned about the Gantt chart, if you steer the discussion toward the third deliverable due eight months from now, then you are not being agile. If you don't trust your workers, we can use Scrum as a method and sell that we are agile. But we are not. Gantt charts are a magnificent visualization tool for many purposes, extremely useful until they come to be considered more than that. Unfortunately, they are taken as a kind of Immutable law of how things will go in the project.

Concerning planning, many agile methodologies use somewhat surprising mechanisms. For example, cards in a Fibonacci sequence (1, 2, 3, 5, 8, 13 ...) are used in a sort of poker game. Why in this bizarre way? Everything has a scientific basis. This is done for the same reason that we fail with long schedules: humans are lousy at evaluating absolute quantities and long-term durations. So, we find it problematic to assign a quantity of work, without further ado, to a task. It is difficult for us to know how much an object measures. Some people with Asperger's are precise at determining magnitudes, but these are exceptions. Humans are good at comparing, though. We can see that one object is larger than another. In music, it is the same: most people—even many professionals—sing by tone reference. Before starting to sing, they request the initial tone—which usually coincides with the key on which a song is going to be performed—and from there, they are guided by distances between tones and semitones. Very few people can hear an isolated frequency and interpret that sound to be a flat 4 sun; this is called "absolute pitch." Very few humans have absolute pitch, absolute sight, or anything absolute: the normal thing is that we are clumsy when estimating magnitudes.

The classic long-term planning also fails for another well-established reason, but rarely admitted: plans are routinely made using the preceding year's numbers as a template, increasing or decreasing the figures by a coefficient. Letting the past influence our vision of the future is wrong. Being stubborn about not wanting to update the plans mid-year, no matter what is going on out there, is bad too. This way of planning on engraving stone penalizes innovation for a phenomenon known as "marginal thinking." In the face of an investment decision, if you have to choose between something without a traceable past or something that has a visible income trend—a continuation choice—the second will always prevail. The second option usually has its fixed costs covered, so the investment is covered thanks to previous marginal gains. An investment decision with little apparent risk to the executor. The problem is that these types of decisions, when taken continuously for long enough, lead to a situation where there are only one or few escape routes. On the other hand, for competitors, especially newcomers, the blind option is the only one possible—and, as we have seen, their entry costs are getting smaller. Hence comes the disruption.

What does an agile business culture look like instead? Start with a foundation of trust. People are free to invent and explore solutions independently. The world does not end if something fails. Failure and learning about errors are encouraged because experimentation and failure are prerequisites for innovation. For this to happen and work, the center of gravity of as many decisions as possible must be lowered to the lowest level of the organization. If people feel that power and responsibility, and if they feel that they can do what they consider best for the company, they will appropriate their products as their own.

Without this precondition, an enterprise cannot become agile simply by implementing Scrum, or whatever method.

If the upper management expects agile methodologies to solve all their problems, it will be a major disappointment. Agile is a movement and a set of methods that push us to change our mentality. We must maintain coherence between what we do and what we think, and keep these two dimensions in sync as we go along. Many organizations ignore this. In those cases, hierarchical control and micromanagement come to prevail and addiction to long-term planning never stops. The influence of the agile product owner is greatly limited. When something goes wrong, the board of directors requires an explanation and desperately searches for a responsible party.

No framework can fix this on its own.

Along with this old corporate culture, traditional budget management also persists. Capital is allocated annually, by each department, and is generally immutable. In the best of scenarios, an unforeseen business case is approved when the steering committee is persuaded to get a good return on investment in three to five years. The standard scenario is to still request deliveries in a particular time and within a definite budget, both fixed. Getting a budget for extensions is difficult within the same financial year, even if all deadlines and deliveries have been previously met. The scope of the project is predefined and materializes in an expected product that the board of directors has in mind, without first asking the customer what they think. Later, the pressure will increase when the deadline is no longer feasible due to mid-project changes.

Under this scheme, the malleability of agile projects becomes impossible.

Agile projects invert the equation. Instead of having a fixed scope, to which the resources and the plan are adapting, a work team and a rate of delivery remain stable, leaving open the scope and, therefore, the final cost:

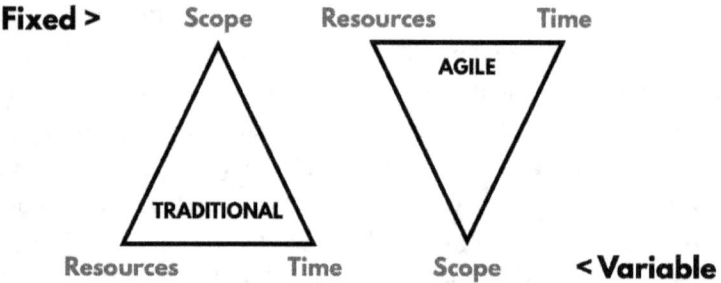

Figure 24: How agile and traditional projects deal with the classic triple restriction of project management.

Combining both agile and waterfall methodologies creates a scenario that is even less efficient and effective than the previous scenario. This phenomenon is known as *water-scrum-fall*: embedding Scrum in a company that previously used waterfall. Firms that run long, linear processes begin to embed agile teams within that same organization. In short: we have budgets assigned once a year. There are staged project-approval processes, which require a prior financial study, which then must be accepted by a committee that meets, say, every quarter. Then, procurement processes follow different times and realities in parallel to the work teams. Project offices manage both waterfall and agile teams at the same time. In this context, it is of little importance to having an agile development process, if sooner or later it will be stopped by a process that will take months. This is the second major problem of agile adoptions by large companies.

Therefore, for some years now, scaled Agile frameworks have become fashionable in large corporations. These frameworks provide a series of guidelines for moving from disconnected teams to organizations that work together nimbly. They arm us with protocols to group projects into programs; assign them a sponsor, budget, and staff; decide which projects are best; improve strategic communication between areas; and so on. Most of them provide a short phase of preparation before starting to develop. It defines requirements, managers, and planned deliveries. Even if we don't want to fully adopt a framework, grabbing certain artifacts can be an excellent idea. Most of these are descriptive frameworks, describing what we should do, as opposed to prescriptive, that show exactly how it should be done, documented to the millimeter. It can be a foothold if you are one the kind of person who cooks following the recipe point by point and gram for gram. Let us see some examples.

The DAD (Disciplined Agile Delivery) framework is a hybrid framework created by former IBM workers. It combines practices drawn from various methodologies (mainly Scrum, XP, Kanban, and Lean), but adding the particularity of including concepts associated with the DevOps discipline for continuous delivery and deployment. What is DevOps? It is the Agile mindset applied to the relationship between the development and the operations departments. DevOps's primary mission is to automate and monitor all steps from software construction, testing, and deployment through to infrastructure deployment and management. To do this, it introduces several Agile concepts, such as shorter development cycles, greater frequency of implementation, close alignment with business objectives, etc.

LeSS is the acronym for "Large Scale Scrum." It is literally what it advertises: taking the activities of a Scrum team to a corporate level. LeSS assumes that some projects are working on similar themes (programs) and takes a representative from each of them to form synchronization teams, or teams of teams. They then follow various Scrum ceremonies (such as daily standup or retrospectives) as if they were an ordinary work team, whose aim is keeping the criteria of each of the projects unified from there to the entire organization. Among the options is one of the most flexible and simple frameworks.

On the opposite side, we find the SAFe framework, a free knowledge base applying Lean Management and Agile best practices with patterns tested by their authors and frequently renewed. It follows a versioning system, as in software. At the moment of writing this book, they are in version 5.0. It's free; this means that all the knowledge is available on its website, although it is possible to acquire the book, become certified, or seek expert help by hiring a consultant. Something very interesting about SAFe is that it provides several levels in which we can apply its practices:

- A Team level, which allows us to group several teams working under Scrum or XP in an orderly way; and a program level, where the efforts of multiple agile teams are integrated to offer value to the company in an Agile Release Train (ART). These two levels are since version 5.0 joined into an "Essential" layer;
- A Portfolio level, where the programs are aligned with the business strategies and your investment intention and concepts such as agile budgeting begin to appear;
- A Large Solution level, specifically for building complex solutions;
- A Full-SAFe configuration, combining them all.

Which choice is the best? It always depends. You can steal from several.

If your organization has never heard of agility, don't worry. It is not too late. Surveys show that Agile adoption didn't become important at the corporate level until after 2010. But dozens of statistical studies show that Agile improves the quality, speed, and market adjustment of the product, and avoids the failure of many projects and the corresponding associated expense.

The two problems explained on these pages should be considered, though. Being agile at the organizational level does not mean having many teams working with methodologies, it means being reconverted at the organizational level, from the way we plan, how we invest in our projects, what we value, and where we place resources. And the second consideration: culture is everything. If it is not adopted with conviction, if we apply a cosmetic Agile, if we object to what the frames recommend, the transformation will not take place successfully. Every year, the *Gallup State of the Workplace* report is published, a working environment survey conducted in over 140 countries and with over 100,000 respondents. The conclusions could not be more pessimistic: only 34% of those questioned stated that they felt committed to their company while 53% were not engaged. And what is worse: 13% declared themselves "actively not committed." That is, they are on the verge of industrial sabotage. It is time to do something about it.

The range and variety of agility are enormous. Unfortunately, most companies only scratch the surface and just implement some Scrum artifacts, like working on sprints, retrospectives, or writing user stories and tasks. Scrum works excellently at developing solutions but is far from offering a complete solution. It omits high-level issues that organizations need to address. With Scrum, we can reach a viable product in less time than was acceptable with a waterfall approach, but that is not enough. The lack of reaction to market changes that many companies suffer from is due to architectural reasons and not just individual productivity. Although the introduction of agile philosophies at the team level is very beneficial, especially in development teams, its path is insufficient. Further barriers soon appear, limiting people's ability to exploit their maximum work potential. Once in production, other practices such as DevOps allow technology departments to deploy software with greater frequency, stability and quality, and fewer failures. Silos are the product of closed hierarchical structures and inflexible human resource policies. New ways of allocating human resources to the tasks that most affect the organization at all times must be put into practice, of thinking about more dynamic ways of assigning budgets, of prioritizing projects by consensus between departments. Otherwise, the result is that each silo executes its projects with its economic resources and its personnel independently, without coordination with the rest of the company. The beneficial decentralization of decision-making must come from prior coordination of the organization's major objectives and not from silos formed by not working together. This is where the scaling frames come into play, which helps us step by step to increase the philosophy at the company level.

EPILOGUE

ETHICS AND POLITICS

9

> *"If machines produce everything we need, the outcome will depend on how things are distributed. Everyone can enjoy a life of luxurious leisure if the machine-produced wealth is shared, or most people can end up miserably poor if the machine-owners successfully lobby against wealth redistribution. So far, the trend seems to be toward the second option, with technology driving ever-increasing inequality."*
> —Stephen Hawking, 1942–2018

These were Stephen Hawking's last words on an Internet forum, during a Q&A session organized by the news aggregator Reddit, on October 15, 2015, two-and-a-half years before he died.[1]

Our worries about unknown technologies are groundless. Strong Artificial Intelligence will not come soon and perhaps it will never be what we understand by the human intellect: self-conscious, sensitive, wise. Robots will not steal our work. If they do, it will be the most tedious, manual, and repetitive portion of it—the kind we will want to give away. And if one day they completely replace us, it will open the doors to a world without labor, in which to express ourselves through the arts, leisure, philosophy, and individual relationships. Machines cannot think, cannot solve problems on their own, and consequently cannot cause them. If we were to ask ourselves today if the Internet has created jobs and wealth, the general answer would be yes. The Internet, however, generated and continues to generate fears, security holes, unwanted leaks, psychoses of parents who ignore what their children are up to on the networks and, of course, have made thousands, millions of obsolete chores disappear. The Internet took over much of the postal system; the cell phone replaced the landline phone; the landline phone supplanted the telegraph. The show goes on.

The exchange of misgivings between workers and technology is as old as the mountains. An anecdote, attributed among others to Milton Friedman, tells that the economist visited a certain Asian country and observed many laborers in public works, using only shovels.

"Why don't they have bulldozers?" He asked.

"It is a job creation program."

"Well, in that case, why don't you give them spoons?"

The situation is absurd, but there is a genuine background. We keep hardcoded the need to create jobs and at the same time, we base our concept of employment on the fact of being busy. At work, those who look the busiest seem more essential. "They must be doing very important things," we reflect.

Rather than blaming the bulldozer or laughing at the use of shovels, it would be appropriate to turn to our inconsistency. Why do we need to create employment? To generate well-being for the people, to have a means of survival. What would happen if we could provide that not through our work, but the machines? This idea is less revolutionary than it seems, but we get blocked by a mental—not technological—wall. With today's technology, the percentage of automated production could be much higher than it is. We are stopped by mechanisms similar to Friedman's anecdote. What if, in the age of robotics and automation, generating more jobs was a problem, not a solution? Perhaps it is time to consider ideas such as the reduction of working hours or the universal basic income.

But before projecting those pathways, we may address a more immediate issue: Hawking's question. How will the fruits of that new workforce be distributed? There has been a lack of synchronization between wage growth and labor productivity for a long time. In economics, it is known as the "Great Decoupling." According to the measuring method used, this disharmony is of a greater or lesser degree, but few scholars argue against it. Wages grow less than the output, also the ratio of the number of hours worked to productivity we had a century ago. The total yield is increasing thanks to automation, but that does not mean that we work less. The demand for employment is fewer than the supply — the active population accepts what they did not previously accept. Precariousness spreads. Employees perceive this and look for the devil on the chip. But it is a political problem, not a technological one. Technology must be there to serve us, for it to do things. He is our servant and not the other way around. It is a quagmire that should be resolved in parliaments, in gatherings, and the streets; at the ballot box or in the assemblies, and not through the defenestration of machinery. Under the North American New Deal — and in all countries during the postwar period — the decline in inequality, the improvement of workers' conditions, and the rapid adoption of new tooling in factories were harmoniously combined. It's possible.

In case the income distribution problem was not enough, we encounter the machine-worker relationship dilemma. We are used to perceiving technology as a complement to accelerate productivity: a peasant digging with his hands is slower than one digging with a shovel, and this slower than another mounted on a bulldozer. What will happen at the moment when there will be autonomous bulldozers? The machine-worker dialogue will disappear. The tool will be a laborer itself. This, which had occurred to a lesser extent during each technological metamorphosis throughout history, will take place more than ever with Artificial Intelligence. Automation possibilities will expand beyond the classic, repetitive, and predictable manufacturing tasks of the last century. Machines may not be humanly intelligent, but they will know how to do much more stuff than before. Further than ever, we will return to the "creative destruction," a term coined by the sociologist Werner Sombart and popularized by the economist Joseph Schumpeter in the 1940s. It captures the phenomenon by which innovation replaces traditional forms of use and production by setting up novel processes, markets, or objects. By introducing new actors, those whom they take over are destroyed. Cars replaced horses, but drivers and carriers remained. It is worth asking: What will happen to our current economic system, in an automated world in which creative destruction does not give way to new jobs (where it is destructive, and not creative)? What will happen when new markets, those blue oceans on which the old disruptive entrepreneurs depended, no longer exist because there will be no demand to satisfy them since there will be fewer jobs, and therefore less aggregate salary? What will happen when the fiscal crisis of the nations, derived from the shortage of income because of the enormous unemployment, does not allow governments to substitute private for public sector employment, as countries that have adopted a Keynesian policy have traditionally done in bad streaks? What will happen when the lack of real growth cannot be hidden by the economy's financialization, a tendency in developed states in recent decades? What will happen when there is no escape for the middle class who avoided the automation of factory work through degrees and diplomas that offered an intellectual value? When the inflationary escalation of higher education cannot be fed, artificial intelligence will engulf it. What will happen then?

Robots do not cause problems, we do. The panic to automation is that of the intimate wisdom of someone who knows that it has not always been used for the common good. We recognize that there is a greater probability that we will not know how to manage this new knowledge, that it will take on a life of its own, and terminate us. As humans, we fear what we cannot understand, but we must not abstract from the fundamental challenges that will affect us. Although we are not interested in technology, nor do we work in a corporation, or want to start a company, nor are we going to implement any digital transformation project—the discussion is on the table. Specific questions to ask ourselves, for example: should robots pay taxes?

Bill Gates thinks so. He stated this in an interview with Quartz in February 2017.[2] This would be a tax on the capital employed by companies to use them and would help correct the long-term increase in capital income over labor income. Robots, let us not forget, are industrial capital. Hence Hawking's sentence is better understood. The effective rate of corporation tax in the United States has not stopped decreasing since 1950, when it was placed near 50%. Since the 2008 crisis, it has fallen below 20% and after Donald Trump's victory, a brutal decrease in the statutory rate was carried out from 35% to 21%, with the Tax Cuts and Jobs Act of 2017. Likewise, the aggregated corporate tax contribution to US GDP has decreased. In contrast, the main personal spending tax, VAT, has been increasing steadily in all countries since its inception a few decades ago. The OECD average in early 2018 was 19.2%.[3] A study by the Taxation and Economics Policy estimates that Amazon paid nothing in federal taxes in 2018. The same as in 2017. On the contrary: it has been returned.[4]

What does this mean? People are giving in greater proportion to tax revenues, through taxes on wages and expenses. Companies are contributing less through income taxes, even though they use transportation and financial infrastructure and benefit their employees through education and public health care. At the same time, salaries are not increasing in line with rising productivity. The capital income grows more than labor income. The result has been an increase in inequality worldwide.

Very interesting aspects of innovation are marred by questions of social justice. A good example is the gig economy, a shared or collaborative economy that we talked about in the first chapters. Some authors point out that some of their successes are based on skipping the existing regulation— be it sectoral or labor, as in the case of companies that treat their employees as independent workers, alleviating their labor responsibilities and social charges. They believe they will start a global spiral of dwindling wage costs when they connect and expand to developing countries and they all compete to do the job for less salary. And they remember that many professions are regulated for good reason. Will we admit the disruption of startups to the public health or judicial system? This is not a technological discussion, nor a business model. It opens the doors to a perhaps undesirable social transformation. After the 2008 economic crisis, which continued to be felt for more than a decade in some countries, we still find people that survive by monetizing their discomfort. They live almost daily with a stranger, someone new every three days, in exchange for paying their rent. They are eventual guests to assets they don't own. Some authors cunningly allude that the term "shared economy" is inappropriate, since the moment an economic transaction comes into play, *sharing* ceases. A Harvard Business Review article in 2015 suggested using the term "access economy" instead.[5] Other sides of this economy, such as ecological profit, are true and beneficial. Regardless of whether it is a commercial sale or a friendly sale, the efficiency in the management and consumption of resources improves. But they are contingent on everyone participating in them. The virtues of austerity are such if they all share the same values. If the industrial giants do not fold to accept a more austere and less polluted planet, little of what we do individually will do.

The questions that shine through are: how many wonders served by our technological development will we be missing because of not knowing how to share? How much will we be wasting the virtues of our Promethean fire, the ability to innovate for our benefit, for not wanting to divide it?

Should we demonize firms like Amazon for evading taxes, or Airbnb and Uber for generating an economy like the one described in the previous paragraph? I think not. Should we demonize our political representatives, and by extension ourselves, who are the ones who vote for them, for allowing a system that makes a huge chunk of society precarious despite the abundance of knowledge and technological plethora that we witness? I think so. Companies do not rule the world, in theory; if they do, it is because we citizens allow it. Governments have responsibilities to their nationals. The good ones try to respond equitably to all residents—even corrupt governments benefit a proportion of the population and fear the general public. Corporations do not need to do any of this. Their mission is to make money. It is the only thing they have to worry about. Social change will hardly come from them.

The prime reason behind the VOC's success was the sponsorship and protection of the Dutch state—more specifically, the military. Centralization, promoted by the authority, also allowed spice prices to always have a wide margin, which is typical of monopolies. This act set them apart from the British, who had several private initiatives trading in Indonesia on their own. The VOC not only made

Ethics and politics

war on the Portuguese, British, and Spanish: it committed atrocities against the settlers, such as the Banda Islands massacre. Slave labor enriched them. Jan Pieterszoon Coen, Governor-General, wrote: "There is no trade without war, nor war without trade." We should remember that the VOC had military support from the government *to trade nutmeg and pepper*. What is valuable is not constant, it changes in time. It should make us reflect on what we value personally and as a society. Were all those deaths worth it to bring cinnamon to Europe?

Companies that emerged from the Chinese economic experiment, such as Huawei, resemble the VOC. Without the explicit support of the government, giants like Amazon, Google, or Facebook have become determining factors for markets, civil society, or democracy. This is dangerous. Google can destroy small businesses just by changing the algorithm of its ads. Amazon with the bookstores. Facebook can do it with democracies—it is difficult to predict where these companies are going. But sovereign states increasingly depend on these companies. They keep their information in their clouds; they handle more complete and immediate knowledge than intelligence services. The VOC should serve as an example of what happens when corporations become more powerful than the nations themselves.

And, perhaps, we should rethink some of our indicators to measure if things are going well.

THE PROBLEM OF THE MASSES

The opening bars of *The Revolt of the Masses* by Ortega y Gasset are curiously familiar from today's standpoint of oceans of data:

What we talk about when we talk about innovation | 251

> *There is one fact which, whether for good or ill, is of utmost importance in the public life of Europe at the present moment. This fact is the accession of the masses to complete social power. (...) Agglomeration, fullness, was not frequent before. Why then is it now? The components of the multitudes around us have not sprung from nothing. Approximately the same number of people existed fifteen years ago. Indeed, after the war, it might seem natural that their number should be less. Nevertheless, it is here we come up against the first important point. The individuals who made up these multitudes existed, but not qua multitude. Scattered about the world in small groups, or solitary, they lived a life, to all appearances, divergent, dissociate, apart.*

The theory of masses in Ortega y Gasset's case does not refer so much to the number of people as to their behavior. The crowd enjoys the new anonymity. Enjoy the hustle and bustle. These lines are printed in the interwar period, with Mussolini in power in Italy and the Bolsheviks in Russia. The notion of "social psychology," however, had been used for decades after the works of Wilhelm Wundt, with *Völkerpsychologie*, and notably the publication in 1895 of Gustave Le Bon's work *The crowd: a study of the popular mind*, and Norman Triplett's studies on social facilitation. Social facilitation is the phenomenon by which we usually complete basic tasks better when someone is watching than when we are alone, and the other way around for complicated ones. Ortega y Gasset's essay was published in 1929, at the peak of fascism. During the postwar period, a considerable interest arose in studying the effects of mass behavior on individuals. Various sociological tests abounded in the concepts of group thinking. In 1949, Orwell publishes 1984. In 1951, Arsch's experiments significantly showed the power of conformity in the masses. Volunteers were asked to take a vision test and only one was a real volunteer, while the others were accomplices in the experiment. The exercises were simple (i.e. comparing and deciding which was the shortest line among several) — so easy that, under normal conditions, the error rate was less than one percent. The accomplices failed on purpose, however, and under peer pressure, actual volunteers were induced to name the wrong answer one in three times. In 1967, the Third Wave experiment was carried out; it was a fictional movement created by a California teacher to explain to his students how the Germans could accept the actions of the Nazi regime. The event was later taken to the cinema on several occasions. According to the Third Wave, it only took four days to get out of control and be aborted.

How is that possible? It seems that people lose their judgment when faced with the plurality of the crowd. Psychologist Irving Janis coined the term *group thinking* based on the Newspeak of Orwell's novel. New technologies expand our exposure and the risk of group thinking increases. The philosopher Byung Chul Han delves into the particular homogenizing role of social structures in The Expulsion of the Different. This has led to an in appearance inexplicable paradox: at the time of greatest access to information and knowledge, citizens, the digital mob, seem to be more manipulable than ever. The most absurd fake news is the most quickly shared. And this is just the beginning of the story — advances in neuronal networks will soon produce hyper-realistic audiovisual material, prone to fake news. We will see videos of people doing this, recordings of people saying that, when in reality it will be all a simulation. Given this fact, human responsibility becomes constantly critical. The combination of fake news with a diminished individual capacity to interpret the context of the data and apply critical thinking will further weaken our democracies and plant the idea that technocrats armed with algorithms must govern us because they are the ones who know the most.

We have thus far exposed a macroscopic problem, the exponentialization of mass behavior, and group thinking in a hyper-connected community. But it is possible to appreciate miniaturized versions of the dilemma in more specific contexts, particularly the role of algorithms as homogenizers of society and amplifiers of prejudice. Do you remember the case of artificial intelligence with a taste for chips? Cambridge Analytica? Let us look at James Watson. Watson first proposed, together with Francis Crick, the double-helical structure of the DNA molecule. He was a highly respected scientist in the academic environment. But he has also been criticized for linking sexual behavior and intellect to genetic factors and race, to the point of being forced to sell his Nobel medal. Let us put forward a hypothesis. Suppose we could reliably measure wit and that we measured it across the globe, for the entire world population. Assume also that Watson is right: Black individuals are less intelligent than white people. The data corroborates it. But think about this scenario: this lower intelligence is not due to hereditary race factors, as Watson believes, but to educational, nutritional, and historical circumstances. The fact is, blacks are in the majority in poorer and colonized lands—they populate the poorest deciles in some developed societies. We would face discrimination caused by the ancient dialectic between races, and not by an inherent factor. In such a situation, how would an artificial intelligence system act? It correlates data without discussing the causes. The answer is obvious: oppression would endure. The algorithm understands the lower capacity but does not discern the reason—genetic or historical injustice. Dozens of examples like this can be found in Cathy O'Neill's book Weapons of Math Destruction. The question arises: by surrendering our decisions to algorithms, do we risk delaying or even freezing social progress, since we are anchoring our thinking to the time in which we coded the algorithm?

Not only the manipulated opinions and news that we find on the Internet are harmful; technology influences and transforms us. Children tied to tablets from birth develop motor problems and speech and communication difficulties, beginning to speak at later ages than usual.⁶ Children no longer play or interact in the same way with each other and we still don't know how this will affect our race in a few decades. Some authors, such as Nicholas Carr, argue that technology and the internet are negatively affecting our ability to think. Formerly, taxi drivers had to learn all the streets, which physically developed their hippocampus.⁷ Some claim it is not useful to memorize roads when a system provides them to us, like learning arithmetic can be useless if we have a calculator. Up to what point?

TOWARDS A WORLD OF LEISURE AND ABUNDANCE?

With Manhattan's population density, the entire human race would fit in New Zealand. It is not a quantity problem; it is a consumption issue. Although we may fit in New Zealand, we use half of the habitable land extension—38% counting inhospitable and sterile land—for livestock and agricultural usage.⁸ Fauna and flora ecosystems that we depend upon hardly survive in the still intact portion of the territory. Many of the countries—and not necessarily the richest—live on a diet beyond the physical possibilities of the planet. For example, if we all adopted New Zealand's diet, we would need two planets to produce the necessary amount of food.⁹

Perhaps it is not so much a conflict of consumption as of waste. Europe throws away 280 kilos of food per person every year. Two-thirds are tossed during its production and distribution; one-third by the buyer. In other words, every European wastes a kilo of food every four days. In supermarkets, the reasons for food loss are not only technical: they reject much in good condition because the food does not adhere to the customer's aesthetic standards. In developing countries, wastage occurs to a greater degree on the productive side, due to lack of infrastructure. In North Africa, 16% is discarded by the consumer himself; in Latin America, 11%; in sub-Saharan Africa, only 3%. However, the total debris, counting industrial and personal, does not differ too much: 225 kilos per person and year in Latin America, 215 in North Africa, and central and western Asia. The champions are the Southeast Asia population with only 125 kilos per person per year.[10] At the same time, global obesity levels have skyrocketed since the 1980s.

A majority of everyday products, such as clothing, have become ordinary merchandise. On much of the planet, dressing needs are met at reasonable prices. At the same time, the fashion industry is second only to oil as the major pollutant. This sector is responsible for producing 20% of the wastewater in the world. It produced 92 million tons of waste in 2015 alone, and only around 1% of textile scrap is actually recycled. With current technologies, it would take 12 years to recycle what fast fashion creates in 48 hours.[11] We buy twice as many clothes as just two decades ago, but the way to differentiate itself is no longer shown by the utility, but aspirational aspects such as the brand, exclusivity, or what celebrity promotes it. Remember, utility versus quality. A 20-euro shirt is not useful—that is, it doesn't do its job—very differently than a 200-dollar one, the same as a watch and even a computer, the most common uses of which are to navigate, check email, play games, watch some movie. Modernity has equaled us like never in terms of taste, attitude, and appearance. This is not necessarily bad as long as we learn to manage it. Once we cover our material needs, the desire to appear different, and transcend widens. The biggest endanger for a leisure society in which we can all access basic inputs for a reasonably merry life is precisely that we would not want to accept it. The Gates of Eden will be opened, but we will not want to enter, because too many people go to that bar.

The innovations explained in this book could lead to a greener system. Internet of things or 3D printing will improve production capacities and minimize waste; they will reduce energy consumption and logistics transport, modernizing distribution channels; it will make water irrigation on farms more efficient. But there will never be a technology that alleviates human greed. Material needs are finite; positional, infinite.

In 2002, at the tender age of 86, Jacque Fresco published *The Best That Money Can't Buy*. Fresco describes a utopian planned society where science and technology are applied with humanitarian and environmental sensitivity, offering us a standard of living beyond the spectrum of the imagination. The machines would take full productive control. The human being would surrender to a contemplative, philosophical, and spiritual existence. Fresco aims to design a global civilization that "respects the carrying capacity of the planet and exceeds the monetary market economic system based on scarcity and poverty." Our inefficient system is what causes shortages, according to him. This claim is hardly undeniable in the face of phenomena such as planned obsolescence or the daily waste of clothes or food. But here is the thing: criticism towards his project is mostly economic, not technical. For example, the social planning he proposes collides head-on with Von Mises's economic calculation problem, according to which a correct allocation of resources is not possible without a market. Surprisingly, the utopian character of the project is conferred more by the description of another society than by a technological limitation. The project's keepers seem to be clear about this: the barrier is political, not technological. Perhaps that is why they have titled their latest documentary *The Choice is Ours*:

> The Project recognizes the important connection between global resource mismanagement and problems such as war, climate change, poverty, and hunger. (...) While technology may succeed in gradually alleviating some of these problems, they cannot be resolved by simply addressing symptoms, as we do now, because they are byproducts of a much larger problem. (...) business interests currently require short-term planning and timely returns on investments. For these reasons, in addition to our expanded technical approach, our proposals include an alternative economic model that overcomes these artificial barriers to planetary wellbeing.

In line with the scarcity denial, we find *Abundance*, written by futurists Peter H. Diamandis and Steven Kotler, and published in 2012. For both, it is clear that technological innovation can ease our problems and offer a prosperous future for all, as long as we learn how to share it. From their point of view, the continuous appearance of disastrous statements does not prove the veracity of these stories or, at least, their statistical significance. That is to say: even being true, they do not reflect the complete truth of things. For them, we are addicts of artificial news that oversimplify reality. Do you remember Kahneman and Tversky? We must recall thinking critically at every moment. Diamandis blames the media for its logic of optimizing clicks by writing sensational and misleading headlines (called *clickbaits*) to capture visitors or viewers.

In this optimistic trip, I highly recommend the work of the late Hans Rosling. In his latest book, Factfulness, Rosling describes ten biases that distort our perspective to veil reality and make us incapable of objectively appreciating the world. His TED talks are among the most visited[12] and certainly among the most fun. For Rosling, the situation is clear: The Earth is getting better and not the opposite. A thorough study of quantitative data shows this. But if we let ourselves be carried away by partial data such as news, subjective opinions, anecdotal cases ... we can easily fool ourselves. We must filter out our irrational fears. What happened to your cousin need not be representative of the rest of us.

Taking the observations of Rosling, Diamandis, and Kotler as true, soon we will have sufficient access to food, energy, water, education, and medical care for the majority. Informatics, medicine, and other areas develop at an exponential rate and in a brief space of time will enable what seems impossible now. Advances are spreading everywhere, allowing the poorest to be substantially improving their quality of life. Techno-philanthropists, millionaires, or technology leaders tackle complex issues, such as cleaning the oceans or eradicating diseases. In the section "We are not alone," we learned how recent technologies harbor an amplifying and feedback effect of their benefits. Those networks will increase in value with the participation of the people still excluded, and their collaboration will improve them. Many of the dilemmas that plague us today are related and their solutions cycle and complement each other. Adequate progress in the standard of living and a decrease in infant mortality will halt the seemingly unstoppable rise in the population of developing countries. This workforce will help solve other problems. The world population will stagnate around 11 billion—what will happen then?

A contemplative life like that described by Fresco, leveraged by digital servants, seems utopian. But there are solid arguments to support this vision or at least a more pleasant existence than we have now. Before the industrial revolutions, the vast majority of the earth's inhabitants were devoted to producing what we need to eat. How amazing would it be for an Italian or German of the time to hear that the percentage of farming employment in their nations[13] is it less than 3% now? It would be a utopia, something inconceivable. Even today in Central African countries the percentage of participation in agriculture is around 80%.[14] But in the global aggregate, we have gone from 44% to 28% in just 30 years.[15] What will happen when not only agricultural jobs will be heavily automated? The systems will still require to be maintained, upgraded, and coordinated. Perhaps we will invent a kind of peaceful military service in which the entire residents of a certain age, say 18, would take care of the machines, while the remainder enjoys. Why not? People 18 years old today account for 1%-2% of the world's population. The same proportion as the community dedicated to agriculture in most of the most industrialized regions. The doors will open to the unbelievable life described by Fresco. Will we be able to take advantage of it? Will we want it?

Again, the decision is ours. Technology helps us solve problems. It enables individuals and networks to fix even more complex ones. But technology attacks symptoms, it does not resolve diseases. How we live, as Hawking reminds us, is up to us. Not up to science. We must feel optimistic about the social, technological, and economic future of the human race. Optimism is the mindset that invokes the Muses, inspires imagination, and fuels the courage we need to be more innovative.

If we don't change the world, at least we can change our lives.

IF YOU LIKED THIS BOOK

If you liked this book…

Thank you for reading this book.

I hope that some of the ideas exposed have resonated in your head or you have been able to learn something new. This work is the result of a mixed professional career of fifteen years in the world of software and innovation in all kinds of institutions, from public and private research centers to small companies and multinationals in several countries.

Special thanks to the professional editor of the English version, Sarah Pila. Thanks to all who helped me in the boring and long process of editing this work in its original Spanish version: Ramón Couto, Ramiro Rodríguez, Sergio Viñas, Jaume Sues, Marcos Cuba, Enrri González, and Isabel Cáceres.

If you think it was worth it, you can consider leaving a comment or review where you bought it, be it Amazon or another site, or simply discuss it with your friends.

To contact me directly, you can find me on the social network LinkedIn: www.linkedin.com/in/alvaroperez.

NOTES AND REFERENCES

THE BIRTH OF TRAGEDY

1. See list of countries by firearm related death rate at <https://en.wikipedia.org/wiki/List_of_countries_by_firearm-related_death_rate/>
2. Roberts, V., Kennedy, E.S. (1959) *The Planetary Theory of Ibn al-Shatir*, Isis
3. Clark, G. (2007). *A Farewell to Alms: A Brief Economic History of the World*. Princeton University Press.
4. Lenin, V.I. (March 13, 1913). A "Scientific" system of sweating. Pravda Magazine, No. 60
5. The term comes from a 2011 book that can be consulted for more information on this topic: *Global Action: The Broken Promises Education, Jobs, and Incomes*, by Phillips Brown, Hugh Lauder and David Ashton
6. *What it's like to be an animal?* <https://www.speedofanimals.com/insect?u=m>
7. Marr, B. (21st of March, 2018). *How Much Data Do We Create Every Day? The Mind-Blowing Stats Everyone Should Read.* Forbes. <https://www.forbes.com/sites/bernardmarr/2018/05/21/how-much-data-do-we-create-every-day-the-mind-blowing-stats-everyone-should-read/#1013c73360ba>
8. Schultz, J. (8th of June, 2019). *How Much Data is Created on the Internet Each Day?* Microfocus. <https://blog.microfocus.com/how-much-data-is-created-on-the-internet-each-day>
9. Exactly 293 billion of daily mails, according to <http://www.radicati.com/wp/wp-content/uploads/2017/01/Email-Statistics-Report-2017-2021-Executive-Summary.pdf>

WE ARE NOT ALONE

1. McIntyre, H. (9th of November, 2017). *Americans Are Spending More Time Listening To Music Than Ever Before*. Forbes. <https://www.forbes.com/sites/hughmcintyre/2017/11/09/americans-are-spending-more-time-listening-to-music-than-ever-before/#463fe6342f7f>

2. Küfner, Sabine. (14th of May, 2018). *Clip—La Evolución de la Startup más Exitosa de México*. In Spanish. <https://medium.com/newco-shift-mx/clip-la-evoluci%C3%B3n-de-la-startup-m%C3%A1s-exitosa-de-m%C3%A9xico-d520bdc6ef51>

3. The case is explained at <https://en.wikipedia.org/wiki/United_States_v._Microsoft_Corp>

4. Check <https://developers.google.com/maps/documentation/javascript/adding-a-google-map>

5. Garnet, IDC, Strategy Analytics, Machina Research

6. More information about the platform can be found at <https://www.doctorondemand.com/synapse>

7. His 2016 TEDxTalk is available on <https://www.ted.com/tedx/events/17437/>

8. Pettit, H., and White, C. (28th of March, 2018). *A glimpse into the future? $39 billion high-tech smart city in South Korea turns into a 'Chernobyl-like ghost town' after investment dries up*. The Daily Mail. <https://www.dailymail.co.uk/sciencetech/article-5553001/28-billion-project-dubbed-worlds-Smart-City-turned-Chernobyl-like-ghost-town.html>

9. In Spanish, check: <https://www.wonderware.es/blog/sistema-de-riego-inteligente-de-parques-y-jardines-para-barcelona/>

10. Wikipedia. *List of self-driving car fatalities*. <https://en.wikipedia.org/wiki/List_of_self-driving_car_fatalities>

11. Hawkins, A. (13th of May, 2018). *MIT built a self-driving car that can navigate unmapped country roads*. <https://www.theverge.com/2018/5/13/17340494/mit-self-driving-car-unmapped-country-rural-road>

12. Euronews. *First self-driving race car completes 1.8 kilometre track*. <https://www.euronews.com/2018/07/16/first-self-driving-race-car-completes-1-8-kilometre-track>

13. Smith, L., Lipner, I. (3rd of February, 2011). *Free Pool of IPv4 Address Space Depleted*. Number Resource Organization. <https://www.nro.net/ipv4-free-pool-depleted>

[14] Li, C. (28th of June, 2018). *Maersk - Reinventing the Shipping Industry Using IoT and Blockchain*. Medium. <https://medium.com/harvard-business-school-digital-initiative/maersk-reinventing-the-shipping-industry-using-iot-and-blockchain-f84f74fe84f9>

[15] Reichert, C. (22nd of September, 2016). *Telstra explores blockchain, biometrics to secure smart home IoT devices*. ZDNet. <https://www.zdnet.com/article/telstra-explores-blockchain-biometrics-to-secure-smart-home-iot-devices/>

[16] Chin, C. (20th of December, 2016). *Kouvola Innovation: transforming the logistics industry with blockchain*. IBM blog. <https://www.ibm.com/blogs/internet-of-things/logistics-blockchain/>

[17] Available at <http://research.google.com/archive/gfs.html>

[18] Dead, J., Ghemawat, S. (2004). *MapReduce: Simplified Data Processing on Large Clusters*. <https://ai.google/research/pubs/pub62>

[19] Chang, F., et altri. (2006). *Bigtable: A Distributed Storage System for Structured Data*. <http://static.googleusercontent.com/media/research.google.com/en//archive/bigtable-osdi06.pdf>

[20] O'Donahue, K. (29th of April, 2018). *How to determine height through the skeleton*. <https://sciencing.com/types-forensic-tests-7551951.html>

[21] Rodríguez Cuenca, J. V., (1994) *Análisis e identificación de restos óseos humanos*. In Spanish. <https://foroporlamemoria.info/excavaciones/intro_antropologia_forense/www.colciencias.gov.co/seiaal/documentos/jvrc03c72.htm>

[22] Kosinski, M., Stillwell, D., y Graepel, T. (9 de abril de 2013). *Private traits and attributes are predictable from digital records of human behavior*. PNAS. <https://www.pnas.org/content/110/15/5802>

[23] Vigen, T. (2015). *Spurious correlations*.

A ROBOT TOOK MY JOB AWAY

[1] Silver, A. (19th February, 2015). *Deep Blue's cheating move*. Chessbase. <https://en.chessbase.com/post/deep-blue-s-cheating-move>

[2] Andrews, E. (5th November, 2007). *A Decade After Kasparov's Defeat, Deep Blue Coder Relives Victory*. Wired. <https://

www.wired.com/2007/05/a-decade-after-kasparovs-defeat-deep-blue-coder-relives-victory/>

3 Kubota, T. (15th November, 2017). *Stanford algorithm can diagnose pneumonia better than radiologists.* <https://news.stanford.edu/2017/11/15/algorithm-outperforms-radiologists-diagnosing-pneumonia/>

4 Ver *Daddy's Car* en https://www.youtube.com/watch?v=LSHZ_b05W7o

5 A demo can be viewed at <https://www.youtube.com/watch?v=NLbKajPS9U0>. It is just an animation, but real-time systems work in a similar fashion, with that kind of dashboards.

6 Delckner, J. (13 de marzo de 2019). *This story was not written by a robot*T. Politico.eu <https://www.politico.eu/article/robot-reporters-newsroom-algorithms-artificial-intelligence/>

DIGITAL METAMORPHOSIS

1 *Ruby on Rails Demo*. YouTube. <https://www.youtube.com/watch?v=Gzj723LkRJY>

2 Reeves, M. (4th of June, 2014). *BCG Classics Revisited: The Growth Share Matrix*. <https://www.bcg.com/en-mx/publications/2014/growth-share-matrix-bcg-classics-revisited.aspx>

3 Perna, Laura, et altri. (5th of December, 2013). *The Life Cycle of a Million MOOC Users*. <https://www.gse.upenn.edu/pdf/ahead/perna_ruby_boruch_moocs_dec2013.pdf>

4 Kolowich, S. (13th of December, 2011). *Stanford's open courses raise questions about true value of elite education.* Inside Higher Ed. <https://www.insidehighered.com/news/2011/12/13/stanfords-open-courses-raise-questions-about-true-value-elite-education>

5 *Noam Chomsky on Propaganda - The Big Idea - Interview with Andrew Marr*. YouTube. <https://www.youtube.com/watch?v=GjENnyQupow>

6 El País. (9th of January, 2020). *'Influencers' nocivas para la salud*. In Spanish. <https://elpais.com/sociedad/2020/01/08/actualidad/1578509328_514133.html>

7 The Economist. (17th of January, 2013). *Shape Up*. <https://www.economist.com/special-report/2013/01/17/shape-up>

8 Gallup, Inc. (31st of March, 2011) *Americans' Top Job–Creation Idea: Stop Sending Work Overseas*. <https://news.gallup.com/poll/

9 Moser, Harry. (8th of July, 2019). *Reshoring Was at Record Levels in 2018. Is It Enough?* <https://www.industryweek.com/economy/reshoring-was-record-levels-2018-it-enough>

10 The complete interview can be watched on YouTube. *Tim Cook Discusses Apple's Future in China I Fortune.* <https://www.youtube.com/watch?v=_ng8xQ-SNGc>

11 West, D. And Lansang C. (10th of July, 2018). *Global manufacturing scorecard: How the US compares to 18 other nations*.Brookings. <https://www.brookings.edu/research/global-manufacturing-scorecard-how-the-us-compares-to-18-other-nations/>

12 Beaudry, P., Green, D., Sand, B. (January 2013). *The great reversal in the demand for skill and cognitive tasks.* <https://economics.ubc.ca/files/2013/05/pdf_paper_paul-beaudry-great-reversal.pdf>

13 XtreeE's project portfolio is available at <https://www.xtreee.eu/projects/>

14 Aprecia's website. *About us.* <https://www.aprecia.com/about>

15 *Statement by FDA Commissioner Scott Gottlieb, M.D., on FDA ushering in new era of 3D printing of medical products; provides guidance to manufacturers of medical devices.* (4th of December, 2017). <https://www.fda.gov/news-events/press-announcements/statement-fda-commissioner-scott-gottlieb-md-fda-ushering-new-era-3d-printing-medical-products>

16 Lord, B.(12th of September, 2018). *Bladder grown from 3D bioprinted tissue continues to function after 14 years.* <https://3dprintingindustry.com/news/bladder-grown-from-3d-bioprinted-tissue-continues-to-function-after-14-years-139631/>

17 Cui, H., Nowicki, M., Fisher, J., y Zhang, L. (20th of December, 2016) *3D Bioprinting for Organ Regeneration.* Adv Healthc Materials. <https://www.ncbi.nlm.nih.gov/pmc/articles/PMC5313259>

HOW TO INNOVATE

1. Keeley, L., Pikkel, R., Quinn, B. & Walters, H. (2013). *Ten types of innovation: The discipline of building breakthroughs*.
2. According to *Quote Investigator*, available at <https://quoteinvestigator.com/2018/01/28/smartest/>
3. Gates's thoughts on IQ are available at <https://www.ndtv.com/video/business/news/what-is-your-iq-sir-ndtv-com-surfer-asks-bill-gates-234319>
4. Have a look at <https://ec.europa.eu/programmes/horizon2020/sites/horizon2020/files/h2020_threeyearson_a4_horizontal_2018_web.pdf>
5. The methodology is explained in a book written by Google engineer Jake Napp, available in bookstores and at https://www.thesprintbook.com. It should not be confused with the "sprints" of some agile methodologies, with which it shares many common values. See Knapp, J., Zeratsky, J., & Kowitz, B. (2016). *Sprint: How to solve big problems and test new ideas in just five days*.
6. Ades, Cely et at. (11th of February, 2013). *Implementing Open Innovation: the case of Natura, IBM and Siemens*. <https://www.jotmi.org/index.php/GT/article/view/1249/801<
7. Hart, David. (17 de agosto de 2004). *On the Origins of Google*. <https://www.nsf.gov/discoveries/disc_summ.jsp?cntn_id=100660>
8. *GPS.gov Program Funding*. <https://www.gps.gov/policy/funding/>
9. Regarding lithium batteries, the article published by the same DoE after the awarding of the 2019 Nobel Prize to Stanley Whittingham, John Goodenough and Akira Yoshino is very interesting. Available at <https://www.energy.gov/science/articles/charging-development-lithium-ion-batteries>
10. Check https://www.darpa.mil/about-us/timeline/ipto
11. *Fact check: The public funding of Elon Musk's ventures*. <https://www.wikitribune.com/wt/news/article/70039/>
12. Bregman, R. (12th of July, 2017). *Meet the greatest inventor of all time*. The original article was written in Dutch. <https://decorrespondent.nl/2496/maak-kennis-met-de-grootste-uitvinder-aller-tijden/95958720-53a49cbb> An English translation by *The Guardian* is available at <https://

13   www.theguardian.com/commentisfree/2017/jul/12/phone-state-private-sector-products-investment-innovation>

13   *What I Couldn't Say...* (9th of March, 2010). http://jonathanischwartz.wordpress.com/2010/03/09/good-artists-copy-great-artists-steal/

14   Key, Stephen. (13th of November, 2017). In Today's Market, Do Patents Even Matter? Forbes. <https://www.forbes.com/sites/stephenkey/2017/11/13/in-todays-market-do-patents-even-matter/#33d3b88d56f3>

15   Data is from Linio, in Spanish. <https://expansion.mx/tecnologia/2018/11/20/mexico-ocupa-el-segundo-lugar-en-ventas-de-e-commerce-en-america-latina>

16   Vikas, SN. (4th of December, 2012). *India Has 137 Million Internet Users & 44 Million Smartphone Subscribers.* Medianama. <https://www.medianama.com/2012/12/223-india-has-137-million-internet-users-44-million-smartphone-subscribers-report/>

17   Salim, S. (4th of January, 2019). *How much time do you spend on social media? Research says 142 minutes per day.* Digital Information World. <https://www.digitalinformationworld.com/2019/01/how-much-time-do-people-spend-social-media-infographic.html>

18   Ransbotham, S., *et altri.* (15th of October, 2019). *Winning with AI. Pioneers combine strategy, organizational behavior and technology.* <https://sloanreview.mit.edu/projects/winning-with-ai/?utm_medium=pr&utm_source=release&utm_campaign=airpt2019>

19   Bucy, M., Finlayson, A., Kelly, G., Moye, C. (May, 2016). *The 'how' of transformation.* <https://www.mckinsey.com/industries/retail/our-insights/the-how-of-transformation>

20   Sturt, D., Nordstrom, T. (8th of March, 2018). *10 Shocking Workplace Stats You Need To Know.* Forbes. <https://www.forbes.com/sites/davidsturt/2018/03/08/10-shocking-workplace-stats-you-need-to-know/#3032c6c7f3af>

21   Ibíd.

22   Ibíd.

START AN INNOVATION FACTORY

1. A recommendation for better understanding better the mapping process is the marvelous book *Mapping Experiences: A Complete Guide to Creating Value Through Journeys, Blueprints, and Diagrams*, by James Kalbach, Ed. O'Reilly, published in 2016, although there are many others.
2. Brown, Bruce, and Scott D. Anthony. (June, 2011). *How P&G Tripled Its Innovation Success Rate*. Harvard Business Review.
3. Tobias, J. (13th of April, 2015). *How Citi Used Innovation To Deliver Growth*. <https://www.thestrategygroup.com.au/how-citi-used-innovation-to-deliver-growth>
4. Desmet, D, Shahar, M., Paquette, C. (November, 2015). *Speed and scale: Unlocking digital value in customer journeys*. <https://www.mckinsey.com/business-functions/operations/our-insights/speed-and-scale-unlocking-digital-value-in-customer-journeys>
5. Check <http://www.service-operating-model.co.uk/>
6. Check <https://www.tmforum.org/business-process-framework/>

COMMON WALLS

1. Bridge, D., Paller, K. (29th of August, 2019). *Neural Correlates of Reactivation and Retrieval-Induced Distortion*. Journal of Neuroscience. DOI: <https://doi.org/10.1523/JNEUROSCI.1378-12.2012>
2. The function is known as the Allen curve: <https://en.wikipedia.org/wiki/Allen_curve>
3. Pontefract, Dan. (11th of May, 2015). *What Is Happening at Zappos?* <https://www.forbes.com/sites/danpontefract/2015/05/11/what-is-happening-at-zappos/#65a560584ed8>
4. There are many interesting books written about the evolutionary relationship of the human being with the oral and written tradition, for example *Wired for Story*, by Lisa Cron; or *The Storytelling Animal*, by Jonathan Gottschall.
5. Almeida, P. (26th of October, 2017). *The brain science behind storytelling*. <https://www.mindproberlabs.com/the-brain-

science-behind-storytelling-part1/>
[6] For some interesting related studies, check <https://blog.sleepnumber.com/why-you-get-all-the-best-ideas-as-youre-falling-asleep/>
[7] Jabr, F. (1st of September, 2016). *Q&A: Why a Rested Brain Is More Creative*. Neurological Health. <https://www.scientificamerican.com/article/q-a-why-a-rested-brain-is-more-creative/>
[8] Cheung, B., Chudek, M., Heine, S. *Evidence for a Sensitive Period for Acculturation: Younger Immigrants Report Acculturating at a Faster Rate*. University of British Columbia, <https://www2.psych.ubc.ca/~heine/docs/sensitivewindow.pdf>
[9] Tsimerman, A., Jaffe, E. (October, 2015). *The Impact of Acculturation on Immigrants' Business Ethics*. <https://www.researchgate.net/publication/282654124_The_Impact_of_Acculturation_on_Immigrants'_Business_Ethics_Attitudes>
[10] Berry, Phinney, Sam and Vedder (2006).
[11] McGovern, G. (2010). *The Stranger's Long Neck: How to Deliver What Your Customers Really Want Online*. A&C Black Business Information and Development.
[12] Available here: <https://news.stanford.edu/2005/06/14/jobs-061505/>
[13] Arrington, M. (15th of July, 2006). *Odeo releases Twttr*. <https://techcrunch.com/2006/07/15/is-twttr-interesting/>
[14] Perez, S. (11th of March, 2019). *Twitter's new prototype app 'twttr' launches today*. <https://techcrunch.com/2019/03/11/twitters-new-prototype-app-twttr-launches-today/>

AGILE

[1] Available at <https://www.agilemanifesto.com>
[2] Toyota Motor Corporation: The Toyota Production System — Leaner manufacturing for a greener planet; TMC, Public Affairs Division, Tokyo, 1998

ETHICS AND POLITICS

[1] Hawking's answer can be accessed from this link. You can also navigate from there to the complete question and answer session <https://www.reddit.com/r/science/comments/3nyn5iscience_ama_series_stephen_hawking_ama_answers/cvsdmkv/>

[2] The interview is pay-per-view, fully accessible from <https://qz.com/911968/bill-gates-the-robot-that-takes-your-job-should-pay-taxes>

[3] Available from the OECD tax database <https://www.oecd.org/tax/tax-policy/tax-database/>

[4] Myers, Kristin. (19th of February, 2019). *Amazon will pay $0 in taxes on $11,200,000,000 in profit for 2018*. <https://finance.yahoo.com/news/amazon-taxes-zero-180337770.htm>l

[5] Eckhardt, G., Bardhi, F. (28th of January, 2015). *The Sharing Economy Isn't About Sharint at All*. Harvard Business Review. <https://hbr.org/2015/01/the-sharing-economy-isnt-about-sharing-at-all>

[6] There are countless studies that link excessive time spent watching TV or using a tablet to communication problems and delayed communication development. A couple of recent examples and in both English and Spanish-speaking populations:

American Academy of Pediatrics (2017). Handheld Screen Time Linked with Speech Delays in Young Children. <https://www.healthychildren.org/English/news/Pages/Handheld-Screen-Time-Linked-with-Speech-Delays-in-Young-Children.aspx>.

Duch, Helena et altri. (1st of July, 2013). Association of Screen Time Use and Language Development in Hispanic Toddlers: A Cross-Sectional and Longitudinal Study. <https://journals.sagepub.com/doi/abs/10.1177/0009922813492881>

[7] Several studies on this subject can be consulted from this article:

Jabr, Ferris. (8th of December, 2011). *Cache Cab: Taxi Drivers' Brains Grow to Navigate London's Streets*. Scientific American. <https://www.scientificamerican.com/article/london-taxi-

memory/>
8 The original source is FAO, but a nice infographic synthesis can be found at <https://ourworldindata.org/agricultural-land-by-global-diets>
9 Ibid.
10 Gustavson, J. et altri. (2011). *Global Food Losses and Food Waste*. FAO. <http://www.fao.org/3/mb060e/mb060e.pdf>
11 The data are collected from various sources and summarized in *Fast Fashion Facts*. 7Billion for 7Seas. <https://7billionfor7seas.com/fast-fashion-facts/>
12 Check <https://www.ted.com/speakers/hans_rosling>
13 Our World in Data. Employment in Agriculture. <https://ourworldindata.org/employment-in-agriculture>
14 Ibid.
15 World Bank. *Employment in agriculture (% of total employment)*. <https://data.worldbank.org/indicator/SL.AGR.EMPL.ZS>

Written in Mexico City, 2020.

www.ingramcontent.com/pod-product-compliance
Lightning Source LLC
Chambersburg PA
CBHW052344220526
45465CB00003BA/944